Lecture Notes in Production Engineering

For further volumes:
http://www.springer.com/series/10642

Berend Denkena
Editor

New Production Technologies in Aerospace Industry

Proceedings of the 4th Machining Innovations Conference, Hannover, September 2013

Springer

Editor
Berend Denkena
Institut für Fertigungstechnik und Werkzeugmaschinen
Leibniz Universität Hannover
Garbsen
Germany

ISSN 2194-0525 ISSN 2194-0533 (electronic)
ISBN 978-3-319-01963-5 ISBN 978-3-319-01964-2 (eBook)
DOI 10.1007/978-3-319-01964-2
Springer Cham Heidelberg New York Dordrecht London

Library of Congress Control Number: 2013946936

Printed on acid-free paper

Springer is part of Springer Science+Business Media (www.springer.com)

Preface

The aerospace industry is characterized by a high degree of research intensity and rapid developments. Due to this, aerospace industry has a high strategic importance in the development of innovative technologies.

"The aerospace industry is one of the most successful industries in Europe, but also one of the most challenging," says Hans-Joachim Peters, Chairman of the Machining Innovations Network e.V. and Head of Core Business Unit Part Production, Premium AEROTEC GmbH. "An important contribution to this leadership role is the innovative capacity of the suppliers with their production equipment and technology research," adds Peters. Despite the long timescales of individual projects due to the necessary approvals innovations are continuously implemented. Currently processes with a high resource efficiency but still good quality and increasing productivity, both in products and in production processes is sought after.

The demand for fuel and resource efficient aircrafts and flights is growing fast from a political as well as social viewpoint. This gives new relevance for the research of new materials and processes, which enable the production of safe and economic aircrafts. Innovative materials need adapted manufacturing processes and this in turn has an impact on planning and organization of factories, machine tools and manufacturing processes.

The Institute of Production Engineering and Machine Tools and the Machining Innovations Network e.V. present in 2013 the Machining Innovations Conference *New Production Technologies in Aerospace Industry*. A total of 26 experts from industry and science will report on two half-days in plenary and technical presentations of the latest innovations and trends. The topics of the different sessions are current trends in manufacturing and production technology as well as planning and organization. For the first time in the 13 year history of the conference, scientific presentations are presented with the latest research results on the key topics of the conference in an extra session. The articles belonging to this scientific session are presented within this issue of "Lecture Notes in Production Engineering".

Therefore, I'm looking forward to the conference as a whole and the scientific aspects in particular with presentations on current research, captivating presentations and lively discussions about the various aspects of *New Production Technologies in Aerospace Industry*.

Contents

High Performance Turning of High Temperature Alloys on Multi-Tasking Machine Tools

U. Karagüzel, U. Olgun, E. Uysal, E. Budak and M. Bakkal

Demands of aerospace industry for high performance alloys have been increasing due to their superior thermal and mechanical properties. These properties, on the other hand, decrease the machinability resulting in lower productivity. Conventional machining techniques can be insufficient to provide higher productivity for these cases. Special processes such as turn-milling and rotary turning can be remedy in increasing productivity in these applications. In order to test the performance of these processes, Mori Seiki NTX2000 mill-turn machining center is used. This machine includes nine axes with two chucks, a milling spindle and a turning turret. The milling spindle whose head moves along the X-, Y- and Z-axes and rotates around the B-axis is used to control the inclination angle in turn-milling and rotary turning tests.

Turn-Milling

Turn-milling (Fig. 1) is relatively a new cutting operation which combines two conventional manufacturing processes; turning and milling. This promising technology becomes an alternative to classical turning due to its advantages such as higher productivity and lower cutting temperatures which provide longer tool life. Intermittent characteristics of turn-milling helps maintaining lower cutting temperatures which make higher cutting speeds possible, produce smaller chips and reduce cutting forces. Parts with large diameters which cannot be turned at high speeds can be machined with increased productivity using milling tools at high rotational speeds. Furthermore, in turn-milling cutting forces applied on the part

U. Karagüzel (✉) · M. Bakkal
Department of Mechanical Engineering, Istanbul Technical University, Istanbul, Turkey
e-mail: ukaraguzel@yahoo.com

U. Olgun · E. Uysal · E. Budak
Manufacturing Research Laboratory, Sabanci University, Istanbul, Turkey

B. Denkena (ed.), *New Production Technologies in Aerospace Industry*,
Lecture Notes in Production Engineering, DOI: 10.1007/978-3-319-01964-2_1,

are reduced reducing deformations. Therefore, turn-milling had the potential to offer increased productivity and quality especially in machining of critical parts.

Tool wear

A cutting cycle in turn milling includes cutting and non-cutting periods which allows cutting edges to cool down reducing diffusive tool wear. In order to investigate this, tool wear experiments were conducted on a Mori Seiki NTX2000 multi-tasking machine. A 32 mm diameter Seco micro turbo 217.69-03 milling tool with three cutting teeth was used in the experiments. The inserts were F40 M grade which is recommended for machining super alloys. The work piece materials were Inconel 718 and Waspaloy which are commonly used in gas turbine hot sections. The cutting conditions used in experiments are as follows: 45 m/min cutting speed, 0.4 mm/rev feed, 0.2 mm depth of cut and 8 mm feed per work piece revolution. Flank wear land on the cutting tools was measured at regular intervals using a microscope as shown in Figs. 1 and 2.

Figure 2 shows tool wear measurement results in machining of Inconel 718 with conventional turning using carbide tooling and turn-milling for different cooling conditions according to the tool life per cutting edge considering the fact that the milling tool has three cutting teeth. Overall, it can be concluded from these results that tool life is improved substantially in turn-milling. Turn-milling, even under dry cutting conditions, provides about 50 % times higher tool life compared to wet conventional turning. It can clearly be seen that coolant has a significant effect on tool wear in turn-milling. For a carbide tool life criteria of 0.4 mm of flank wear, turn-milling with coolant (either conventional or MQL, although MQL seems to yield better life at higher speeds) provides about 3–3.5 times higher life compared to conventional turning.

Waspaloy is another important material in aerospace engineering. Figure 3 shows comparison between conventional turning under dry conditions and turn-milling with wet and MQL cooling for the tool life per cutting edge as the cutting

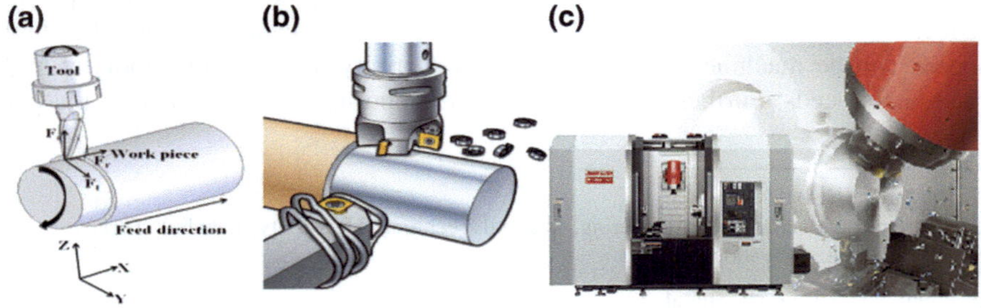

Fig. 1 **a** Turn-milling operation. **b** Chip disposal compared to conventional turning (Sandvik). **c** High performance turn-milling (Mori Seiki)

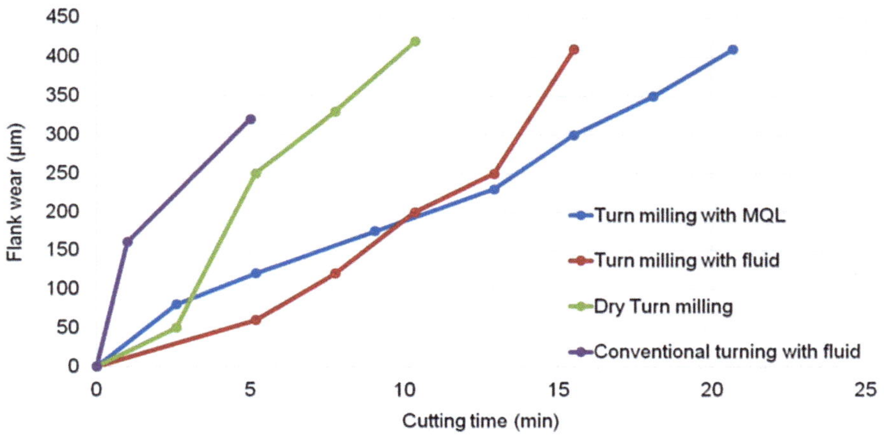

Fig. 2 Tool wear results of Inconel 718 for different cutting conditions

inserts immediately broke in dry turn-milling of waspalloy. One can see from Fig. 3 that tool life is increased substantially (up to 30 times) by turn-milling per cutting edge.

MRR (Material Removal Rate) Optimization

High MRR (Material Removal Rate) is possible in turn-milling but it may cause increased form errors in finishing as shown in Fig. 4. Turn-milling process does not produce an ideal circle. Since in turn-milling cutting tool and work piece rotate simultaneously, the resulting machined part cross section is a polygon as shown in

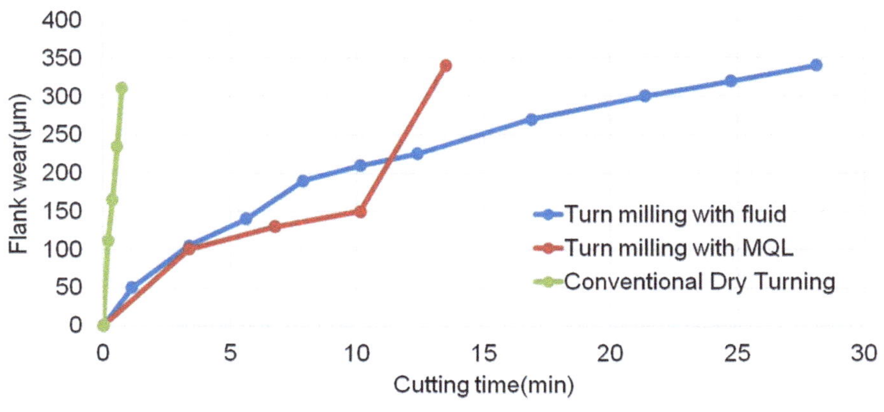

Fig. 3 Tool wear comparison between turn milling and conventional turning of Waspaloy

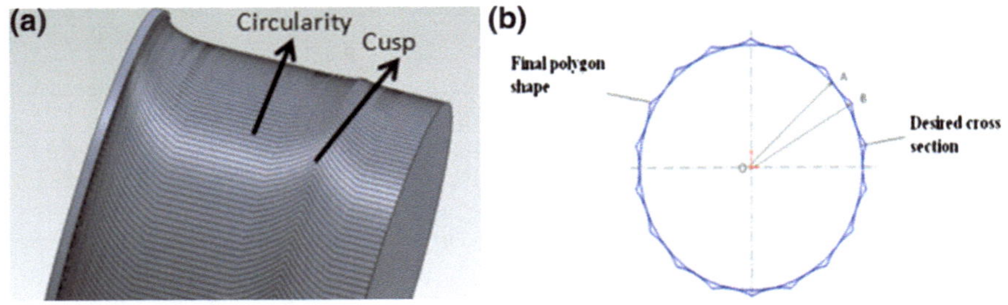

Fig. 4 **a** Form errors in turn-milling. **b** Circularity error (*cross section view*)

Fig. 4. If the feed per work piece revolution is increased for higher material removal rates, the cusp height shown in Fig. 4a is also increased.

Equations (1) and (2) explain the circularity error in turn-milling operations. In Fig. 4b, an ideal cross section and the one obtained in turn-milling can be seen. OB-OA in Eq. (2) describes the difference between desired and obtained cross sections.

$$\theta = \frac{2\pi}{z r_n} \tag{1}$$

$$OB - OA = \left(R_w - a_p\right)\left(\frac{1}{\cos\frac{\theta}{2}} - 1\right) \tag{2}$$

where z is the number of cutting teeth, r_n is the tool speed to work piece speed ratio, R_w is the work piece radius and a_p is the depth of cut. Figure 5 shows the effects of cutting conditions on the form error and tool life. As expected the tool life can be improved by decreasing the cutting speed at the cost of increased in circularity error. Figure 5b shows that MRR can be increased using higher feed per work piece revolution but again at the cost of reduced quality.

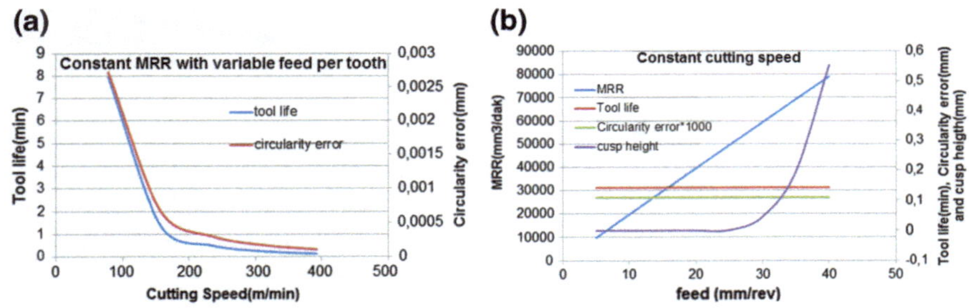

Fig. 5 Effect of cutting conditions on MRR and part quality in turn-milling

Rotary Turning

Rotary turning, which is a specialized turning process, can also be a remedy to improve the machinability and productivity of difficult-to-cut alloys. In this process, the round insert rotates continuously about its own axis during cutting. This tool rotation distributes the generated thermal energy to the whole cutting edge resulting lowered cutting temperatures and uniformly distributed flank wear on cutting edge. There are two types of rotary turning tools which are Self-Propelled Rotary Turning (SPRT) and Actively Driven Rotary Turning (ADRT) tools. In the former one, the tool is rotated by the action of the chip and cutting forces where the tool rotary speed depends on the workpiece geometry and the cutting speed. For the latter one, on the other hand, an external power source is used to rotate the insert. In this type, the tool speed and inclination can be adjusted independently. SPRT and ADRT processes can be seen in Fig. 6a, b, respectively. In this paper, only the ADRT tool performance for various difficult to machine alloys is presented.

The experimental set-up is shown in Fig. 6c. The tests were conducted under dry, flood coolant and MQL conditions with various tool speeds and tool inclination angles. The cutting tool used for ADRT is a carbide insert with multi-layer CVD coating of MT-Ti(C, N) + Al_2O_3 + TiN. It has 25 mm diameter with a chip breaker and 7° clearance angle.

The workpiece materials are Waspaloy, Ti6Al4 V and Inconel 718. 45 m/min cutting speed, 0.1 mm/rev feed and 0.2 mm depth of cut were used in the tests. Three different tool inclination angles (0°, 5°, 15°) and three different tool speeds (10 m/min, 20 m/min, 45 m/min) are tested.

Tool Wear

Tool wear results of ADRT process are presented in this section for various cutting conditions in comparison to conventional turning. In order to properly compare the lives of the stationary turning insert and the rotating one, normalization is

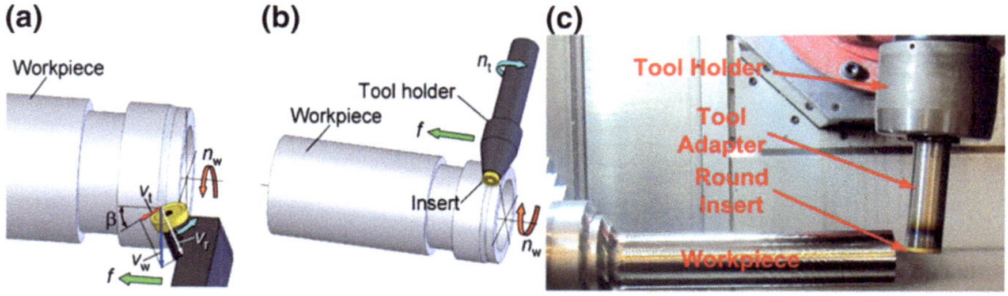

Fig. 6 Types of rotary turning (**a**) SPRT (**b**) ADRT (**c**) ADRT tool position on the mill-turn machine

necessary. This is due to the fact that when the rotating insert reaches the maximum allowable wear all around the cutting edge its useful life is finished and needs to be replaced whereas the stationary insert can be indexed and its unused portions can be used until the whole cutting edge is completely worn. The number of times the stationary insert can be used depends on its contact portion, and thus on the width of cut. In short, the measured tool life for stationary inserts must be multiplied by the number of indexing which is called normalization in this analysis. Figure 7a exhibits the *normalized* tool life test results for different rotary tool speeds on Waspaloy including conventional turning (Vt = 0) results. Cutting test results for Waspaloy indicate that increasing rotary tool speed after a certain range causes higher tool wear rate. At very low and high tool speeds, the tool life is the worst due to thermal effects. At low speeds the contact time between the tool and the workpiece increases whereas at higher speeds the required time for the tool to cool down is inadequate resulting in heat accumulation at the cutting edge as seen in Fig. 7a. For the best condition of rotary turning tool, tool life increases 43 % compared to conventional turning tool, i.e. for Vt = 0 m/min. On the other hand, increasing rotary tool inclination angle (shown in Fig. 6 as β) improves tool life as seen in Fig. 7b. Increasing inclination from 0° to 5° increases tool life 44 %, while a rise from 5° to 15° causes a further increase of up to 49 % in tool life.

Similar results are obtained for Ti6Al4 V cutting tests, i.e. as the tool inclination increases the tool wear rate decreases as shown in Fig. 8a. Figure 8b exhibits the effect of cooling condition on the tool life. Using coolant and MQL (Minimum Quantity Lubrication) improves tool life which can be attributed to reduced thermal shock on the cutting insert and more effective transportation of the cutting fluid to the cutting zone. Interestingly, for AISI 1050 steel cutting tests, dry condition gives the best result which may be explained by reduced thermal fatigue between high temperature (in cut) and low temperature (out of cut) regions.

Inconel 718 cutting tests also indicate that as the tool inclination angle increases, the tool life also increases as shown in Fig. 9a. Figure 9b shows the effect of coolant on the tool life where cutting fluid and MQL improves tool life 47 % and 12 % respectively, compared to dry cutting.

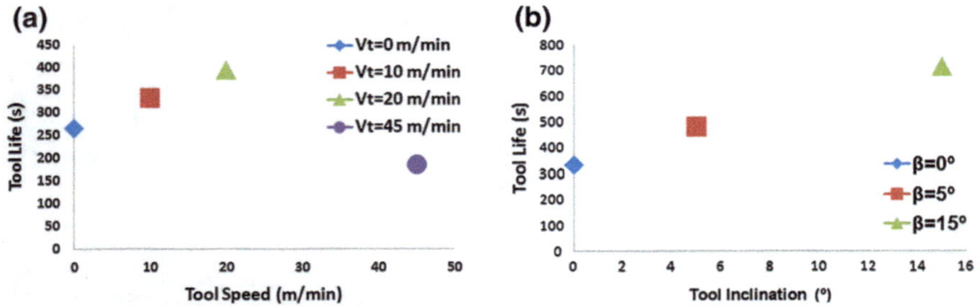

Fig. 7 The variation of tool life with **a** Tool speed. **b** Tool inclination angle for Vt = 10 m/min, cutting with coolant for Waspaloy

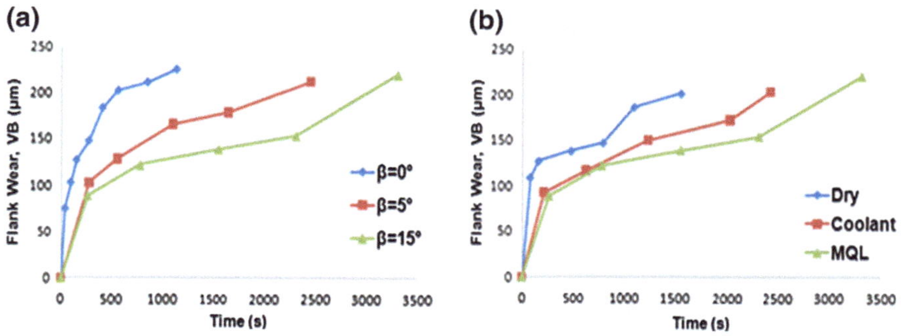

Fig. 8 Tool wear for Vt = 10 m/min **a** Different tool inclination angles for MQL cutting. **b** For different cooling conditions for $\beta = 15°$ for Ti6Al4 V

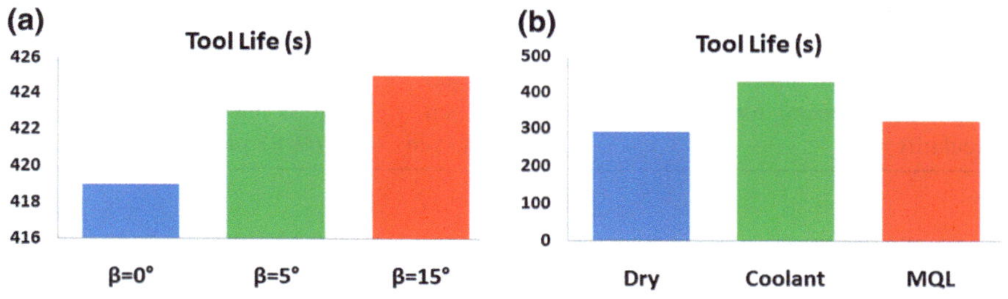

Fig. 9 Variation of tool life **a** With tool inclination angle for MQL, Vt = 10 m/min. **b** With cooling types for $\beta = 5°$, Vt = 20 m/min for Inconel 718

For Waspaloy cutting tests, cutting inserts for different cooling conditions, dry, coolant and MQL are examined under SEM as shown in Fig. 10. In all cases, flank wear is the dominant wear type where crater wear is not observed. High level of workpiece adhesion to the flank face is recognized, especially in dry cutting condition. These deposited materials re-enter the cutting zone due to tool rotation and triggers attrition of the tool material.

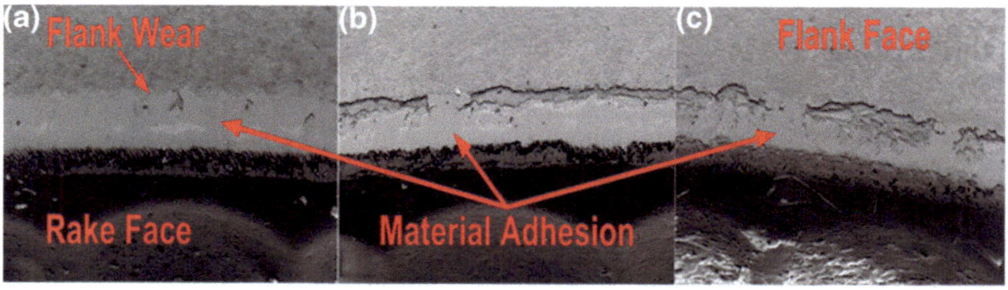

Fig. 10 SEM images of Waspaloy, Vt = 10 m/min, $\beta = 0°$ **a** Dry **b** Coolant **c** MQL conditions

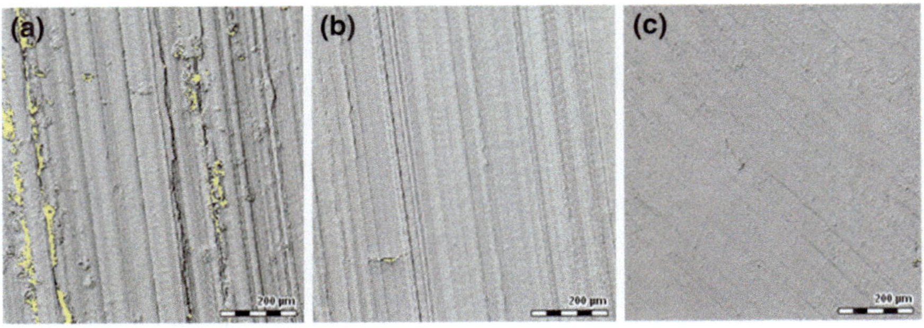

Fig. 11 Surface topography for different inclination angle and cooling conditions on AISI 1050 (**a**) 5°, MQL (**b**) 5°, Coolant (**c**) 0°, Coolant

Table 1 Surface roughness and circularity measurements for various tests

	5°, MQL	5°, Coolant	0°, Coolant
Roughness in feed directory	2.66 µm	1.22 µm	0.92 µm
Roughness in circumferential directory	1.33 µm	1.39 µm	1.44 µm
Circularity	35–43 µm	69–85 µm	29–30 µm

Surface Roughness and Circularity

Surface and form measurements were carried out in order to understand the effects of tool inclination and cooling conditions on the surface roughness and circularity of the workpiece, on test parts made out of AISI 1050 steel. As shown in Fig. 11, for 5° tool inclination, cutting traces on the surface can be seen at an angle to the feed direction due to tool spinning. 0° tool inclination provides better roughness results in the feed direction as shown in Table 1 where coolant improves roughness in the feed direction. MQL results in slightly better roughness in the circumferential direction compared to other cooling conditions. High circularity errors shown in Table 1 indicate that rotary turning process is suitable rough turning of difficult to cut alloys.

Conclusions

Multi-tasking machines offer opportunities to achieve higher machining productivity. Turn milling and rotary turning investigated in this study show that they can provide higher tooling performance in turning of high temperature alloys. Conventional turning is a continuous cutting process which produces high temperatures in cutting of special high temperature alloys. On the other hand, turn

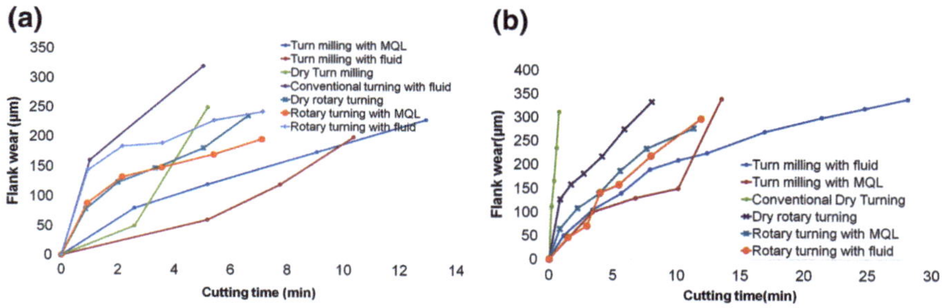

Fig. 12 Tool wear comparisons of turn milling, rotary turning and conventional turning for **a** Inconel 718 and **b** Waspalloy

milling and rotary turning are interrupted cutting operations which offer a chance to reduce cutting tool temperatures and improve tool life.

In this study, turn-milling and rotary turning operations are studied in terms of tool wear and machined part quality under different cutting and cooling conditions. As the summary of the tool wear measurements given in Fig. 12 shows rotary turning and turn milling provide much higher tool life (normalized for number of cutting edge and indexing) compared to conventional turning of Inconel 718 and Waspalloy.

Acknowledgments The support from Mori Seiki Corporation, Pratt and Whitney Canada and Turkish Scientific and Technological Research Foundation (Tubitak—Project 110M522) is appreciated by the authors.

Impact of Clamping Technology on Horizontal and Vertical Process Chain Performance

Roman Kalocsay, Thomas Bergs and Fritz Klocke

Abstract Clamping technology plays a major role in optimization of holistic process chains, determines auxiliary process times, enables process performance and affects workpiece quality. In this study we evaluate three different referencing strategies in horizontal process chains and discuss the effect of fixtures on workpiece dynamics. In order to achieve an optimal process chain performance an optimization of clamping solutions is crucial.

Keywords Clamping · Fixture design · Process chains · References · Damping

Introduction

To achieve robust and economic manufacturing, optimization is not limited to single process steps. A holistic approach must take into account the production process as a whole. It can be structured in two different scopes we call "vertical" and "horizontal process chain". In focus each manufacturing step can be divided into a sequence of tool path planning and machining operations, the "vertical process chain" (Fig. 1). Key factors regarding the optimization of a single operation in "vertical process chains" are machining parameters, tools, tool paths and fixture design amongst others. The "horizontal process chain" addresses the sequence of operations in a larger scope. Horizontal optimization deals with the efficient interconnection of process steps and involves logistics and harmonization of information transfer. Clamping technology can impact both directions of these process chains significantly to improve the overall production process.

R. Kalocsay · T. Bergs · F. Klocke (✉)
Fraunhofer IPT, Steinbachstraße 17, 52074 Aachen, Germany
e-mail: Fritz.Klocke@ipt.fraunhofer.de

B. Denkena (ed.), *New Production Technologies in Aerospace Industry*,
Lecture Notes in Production Engineering, DOI: 10.1007/978-3-319-01964-2_2,

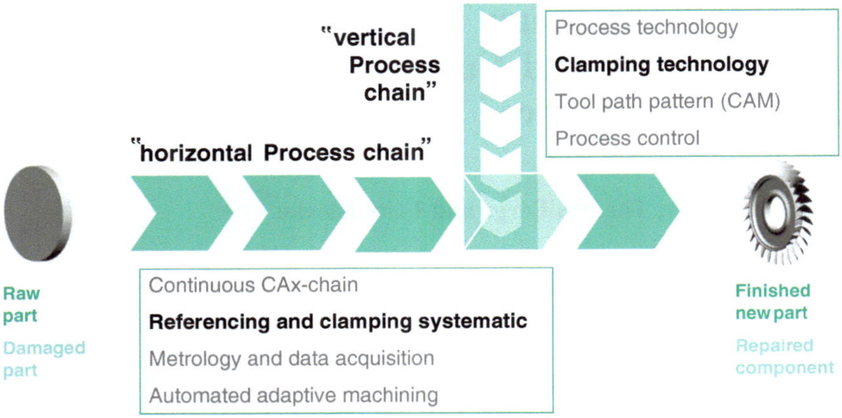

Fig. 1 Horizontal and vertical process chain

A first design theory on clamping devices was given by [1]. Willy [2] contributed a design method for fixtures in automated industry. In [3] active fixture design is outlined and in [4] a method for setup optimization is presented.

References and Positioning in Horizontal Process Chains

The performance of horizontal process chains mainly depends on positioning efforts. Optimization therefore requires dealing with references in clamping. Efficiency aims on data consistency, providing a fast way to establish defined relative positions of machine tools and workpieces in each process step. In the following three different approaches are presented (Fig. 2).

Clamping Device Based References

In this approach the workpiece is positioned at a predefined location in the machine tool. In this case movement is purely force driven using spring based, hydraulic or pneumatic kinematics pressing the workpiece onto rigid locators. To overcome jamming caused by friction before reaching the final stop position, the actors engage sequentially with subsequently increasing forces [5] or subsequently in a pulsewise manner maintaining a steady base load. By this means, a high repeat accuracy with usually less than 10 μm positioning tolerance can be achieved, depending on surface quality and friction deviation. In a second step the achieved workpiece position can be stabilized by attachment of additional supports.

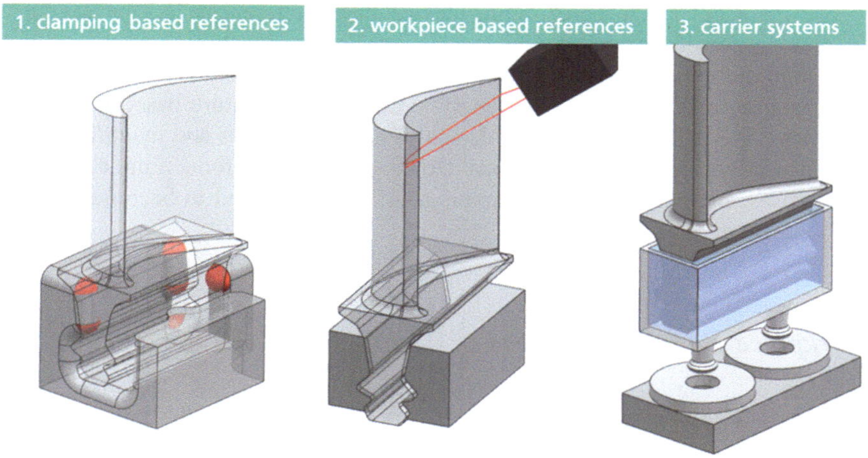

Fig. 2 Three referencing strategies; *left* clamping based referencing by clearly determined workpiece positioning; *middle* workpiece based referencing using data acquisition; *right* carrier system using cast in workpieces combined with a zero point system

In cast or forged workpieces large deviation of the form causes problems in clamping, especially due to uneven burrs at seams. Because of poor surface quality the classical force-driven positioning approach with rigid locators may lead to such large displacements of the center point that, even with reasonable oversize, geometry of the finished workpiece may exceed the boundary of material at machining position. Consequently, the high scrap rate often needs to be reduced by manual adaption efforts. This can be overcome by a centric positioning approach, which uses rigid kinematics (i.e. centric threads, knee levers or gears). It is crucial to keep the elasticity of the moving parts as low as possible. To enhance stiffness a two-stage approach can be followed: Centric positioning is achieved in the first step. In a second step positions of any moving elements are locked based on friction. Additional stability can be achieved by spring-loaded supports that are locked by friction in a third step. Given that kinematics are clearly determined in the positioning step, a repeat accuracy of $<50\ \mu m$ is achievable, depending on stiffness and bearing play of the kinematics.

In both approaches a high repeat accuracy can be achieved, whereas absolute accuracy of positioning might still be poor. High precision locators for exact positioning are prone to dirt and abrasives. Starting from the resulting natural clamped position of the workpiece and adjusting the machining by performing a calibration procedure offers a more robust and consistent approach. To determine the transfer matrix, the measured position of a reference workpiece relative to the coordinate system of the clamping device is compared with the CAD data being the basis for the CAM programming. The reference workpiece represents the average of the clamped parts, having reference geometry attached.

The main advantage of clamping device based references is time saving as referencing steps for each individual workpiece are dispensable. Combination with zero-point systems for setting up the clamping device provides high flexibility in interchanging machine tools. One possible drawback of fixture based referencing approaches is low flexibility as the design of tailored fixtures and manufacturing a reference workpiece can be laborious. However, in serial production or regarding MRO cases, frequent repetitive clamping tasks are expected to become economically efficient.

Workpiece Based References

In the first machining step reference geometry is applied to the raw part, providing a base for subsequent machining operations. The reference is read out at the beginning of each machining step and machining coordinates are adjusted to each individually fixed position. Usually, the reference is generated by removal of material by drilling holes or milling orthogonal faces. Sometimes metal inserts or optical reflectors are used in fiber-composite parts. In future applications the use of RFID technology appears to be promising for non-metal parts as well. Application of the reference geometry speeds up the time-consuming initial calibration procedure. However, referencing times still accrue in each process step for individual parts. Therefore and to provide an overall efficient horizontal process chain it is important to use a fast referencing method as provided by optical or RFID based systems. Up to now most setups are tactile based, offering robust, cheap and accurate but also very slow solutions.

Difficulties with workpiece based references occur when high material removal rates prevent obtaining a final reference or the reference gets lost during processing. Deviations caused by released residual stress or thermal dilatations of large parts also pose a challenge. The combination of a continuous CAx-chain and optical machine-integrated metrology provides a technical solution here. By acquisition of a 3D model of the clamped workpiece and online comparison with the database in real-time, the coordinate transformation for the machining operation can be calculated without additional reference geometry. This approach also allows adaptive reacting to dilatations and workpiece deformations. However, this requires demanding metrology and CAx programming as well as time consuming scanning procedures.

Workpiece based references enable the use of flexible clamping technology with limited repeat accuracy such as construction set fixtures, vices, magnetic plates, multi piston devices (e.g. Matrix^{TM}) or cast in systems without prior positioning steps. Accuracy largely depends on the metrology. The main advantage of this approach is high flexibility, rendering its use in small batch size and being the method of choice for individual production.

Carrier Systems

In this clamping approach a carrier is attached to the workpiece in the first operation. This not only provides the reference for all subsequent operations, but also constitutes the physical fixation regarding machining forces. The carrier remains attached to the workpiece throughout the horizontal process chain and is removed in the last process step, sometimes followed by an additional step to clean the workpiece. The carrier can be mounted locked by friction (e.g. clamps), form locked (e.g. treads) or bonded (e.g. by glue, castings in bismuth or plastic). Connection of the carrier to the machine is usually provided by zero-point systems, giving the advantage of fast and accurate positioning and high stiffness. Zero-point bolts may directly be mounted into the machined workpiece as well. The repeat accuracy of this approach is equal to zero-point systems (± 5 μm); the absolute accuracy depends largely on the first positioning step.

The advantage of this method is the natural preservation of the reference throughout the horizontal process chain. This permits both short positioning times and high precision. The carrier often provides high stiffness and good resistance to process forces and dynamic excitation, especially when bonding connections are used in order to handle complex or thin walled geometries. Drawbacks result from required mounting and dismounting of the carriers. Complex carriers and large process chains can involve high investment costs. Up to now carrier approaches are used in mass production and in cases of very complex workpiece geometry. However, new commercially available zero-point systems with flexible chuck positions open up the opportunity to apply the carrier strategy also to medium production rates and larger production variety.

Workpiece Dynamics Affecting Vertical Process Chains

Vertical process chain performance depends to a large part on fixture design. The most important aspect is convenient handling, influencing clamping times, scrap rate and process quality. Other important goals in vertical process optimization regarding clamping devices are minimizing operating errors, avoiding unwanted workpiece deformations, shortening setup- and clamping times and lowering fixture costs. In the following workpiece dynamics will be focused on to show the impact of fixtures on process performance.

Stiffness and Eigenmodes depend on the valence of bearings and position of supports. Increasing stiffness enables the use of higher federates with larger process forces if static workpiece deflection limits the process performance. However, regarding workpiece vibrations increasing the damping of clamping devices is more efficient than merely increasing the stiffness. Damping in clamping devices can be obtained by using friction, viscoelastic materials (e.g. low order polymer networks), hydraulic systems and other mechanisms. The use of

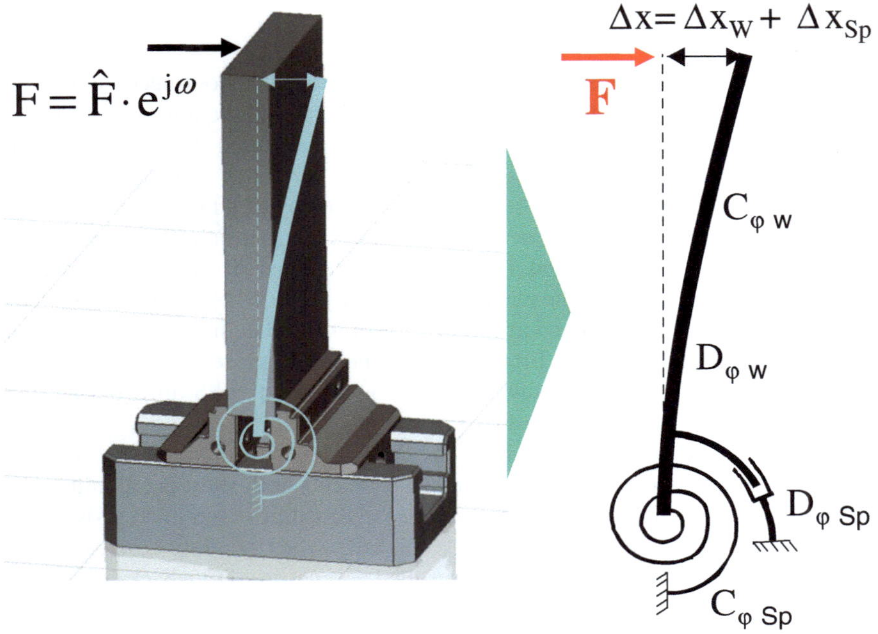

Fig. 3 Effect of damping in clamping devices. *Left* clamped steel plate 150 × 80 × 20 mm, cantilever plate model of clamping devices. *Right* frequency response, $D_w = 0.1$ %, F = 100 N

HIDAMETS (e.g. CuAl12.3Mn5.1) appears to be especially promising because of its relatively high stiffness, suited to be integrated directly into the flux of forces within the fixture design. Figure 3 shows the potential of damping for a cantilever beam setup. In this simulation the resulting resonance amplitudes can be reduced to less than 1 %, if the clamping device is viscoelastically dampened by 30 %. This value represents the maximum achievable damping ratio using high-dampening metals (HIDAMETS).

Compared to classical fixture design the stiffness of dampening clamping devices needs to be slightly reduced to obtain best dampening effects with lowest amplitudes. In Fig. 4 an exemplary fixture design based on CuAlMn is shown having optimized stiffness properties regarding the clamped workpiece, in this case being a 1.6 m steam turbine blade made of steel with momentum free mount at the tip. A periodical force of 100 N excites in the middle at resonance frequency of 122 Hz. The calculated results show that even with low damping ratios resonance amplitudes can be decreased significantly.

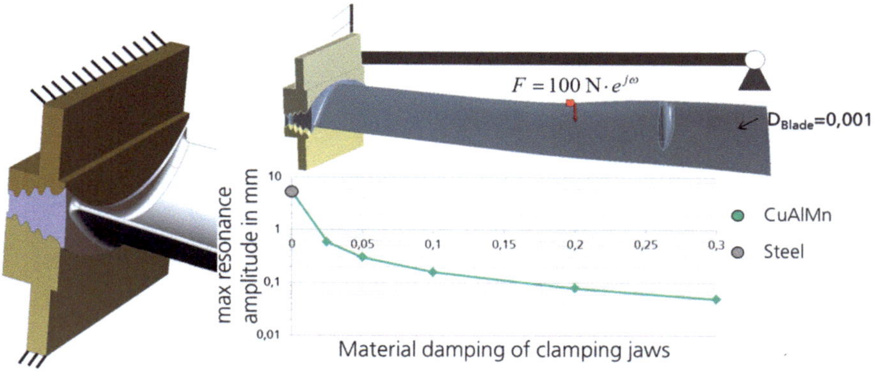

Fig. 4 Resonance amplitudes versus material damping in a fixture for large steam turbine blades

Conclusion

In this study we outline the influence of clamping technology on horizontal and vertical process chains with a focus on referencing and also considering workpiece dynamics. Three different approaches of referencing are discussed and the potential of passive damping in clamping devices for optimization is demonstrated. An optimal strategy in clamping needs to consider a variety of factors as pointed out above and requires solutions tailored to the individual production process. In order to achieve optimal clamping solutions an analysis of process chains is crucial.

This publication contains results of the project "DynaMill", funded by the European Commission within the 7th Framework (FoF.NMP.2012-4).

References

1. Lock, F.: Konzeption und Entwicklung von Vorrichtungssystemen für die spanende Fertigung, PhD Thesis, RWTH Aachen, (1984)
2. Willy, A.: Vorrichtungssysteme für die flexibel automatisierte Montage. PhD Thesis, University of Stuttgart, (1993)
3. Bakker, O.J.: Control metrology and modeling of active fixtures, PhD Thesis, University of Nottingham, (2010)
4. Eichhorn, N.: Rüstoptimierung in der Zerspanung, PhD Thesis, TH Darmstadt, (2012)
5. Raghu, A., Melkote, S.: Analysis of the effects of fixture clamping sequence on part location errors. Mach. Tools Manuf. **44**: 373–382 (2004)

Simulation of the NC Milling Process for the Prediction and Prevention of Chatter

S. Odendahl, R. Joliet, E. Ungemach, A. Zabel, P. Kersting and D. Biermann

Abstract The main goal in the design of milling processes for components in the aerospace industry is the optimization of productivity while maintainig process stability. These two goals can be conflicting, especially if long tools are required, which are particularly susceptible to vibrations. In order to reduce the number of costly experiments, simulation-based approaches can be used to evaluate generated NC programs beforehand. In this paper, a modeling approach for the detailed simulation of engagement conditions, process forces, and dynamic tool behavior is used to detect instable process conditions. Additionally, an algorithm is presented to change the axial immersion in order to avoid regenerative chatter during milling. To demonstrate the effectiveness of the simulation approach and of the compensation strategy, a comparison is shown between experimental and simulated results and between the workpiece generated by the original and the optimized NC programs.

Introduction

The manufacturing of complex structural parts for, for example, air planes, requires precise milling operations with high productivity. On the one hand, these parts often include deep cavities which demand the utilization of long tools, and, on the other hand, large axial and radial immersions are necessary to achieve the desired material removal rates. However, this combination can lead to chatter vibrations which may result in insufficient surface quality or even tool failure [1]. Therefore, the dynamic process behavior is an important aspect of the milling process that should be considered during the process planning phase [2].

S. Odendahl (✉) · R. Joliet · E. Ungemach · A. Zabel · P. Kersting · D. Biermann
Institute of Machining Technology, TU Dortmund University,
Baroper Str. 301, 44227 Dortmund, Germany
e-mail: odendahl@isf.de

B. Denkena (ed.), *New Production Technologies in Aerospace Industry*,
Lecture Notes in Production Engineering, DOI: 10.1007/978-3-319-01964-2_3,

Analytical methods exist for the prediction of process forces and stability for constant engagement conditions [3, 4]. By applying these approaches or by conducting experiments, stability charts can be generated in order to describe the stability limits of a milling process. These charts can then be used to choose a stable combination of process parameter values. However, for the five-axis machining of free-formed surfaces, the engagement conditions may change with each tooth feed. Therefore, the shape of the undeformed chip has to be determined in each simulation step to be able to calculate the machining forces and predict the process dynamics. A simulation system, which performs this analysis, can be applied to detect surface location errors and process instabilities in advance without extensive experimental investigations.

In this paper, a dynamic model for five-axis NC milling processes and the required geometric tool and workpiece models are presented that can be used for a simulation-based chatter detection algorithm. To demonstrate the applicability of the simulation, a local modification algorithm for a given five-axis NC program is applied. In order to prevent chatter, additional path segments are inserted automatically without changing stable sections of the NC program and without the need for additional equipment.

Simulation-Based Analysis of Milling Processes

The simulation-based analysis is conducted using a simulation system which combines different techniques [5] in order to model the cutting forces, the dynamic process behavior, and the surface location error. For the calculation of the milling forces, the shape of the undeformed chip is determined using the Constructive Solid Geometry (CSG) technique [6]. With this technique, simple shapes, e.g., sphere, cylinder, or box, are combined using Boolean operations to describe complex objects. The chip corresponds to the intersection of the tool envelope and the workpiece model prior to the current position in the NC program (Fig. 1a). The latter is generated by subtracting the tool envelopes positioned at each prior tooth feed from the raw workpiece [7]. In order to allow calculating the force progression over the course of individual tooth feeds, each tooth feed is subdivided into multiple simulation steps. The chip shape is scanned by rays cast from the tool axis along the cutting edges for each of those steps (Fig. 1b). The intersections of these rays with the chip shape can then be used to obtain the undeformed chip thickness and calculate the process forces according to the Kienzle equation [8]. In contrast to analytical approaches, this technique is able to model the engagement conditions for any point in time during a five-axis NC program. Since the chip shape is represented as a continuous model, the force calculation can be conducted with arbitrary resolution by choosing the number of rays that are used to scan the chip.

For modeling the dynamic process behavior, the calculated cutting forces are fed into a dynamic model consisting of a system of uncoupled harmonic oscillators

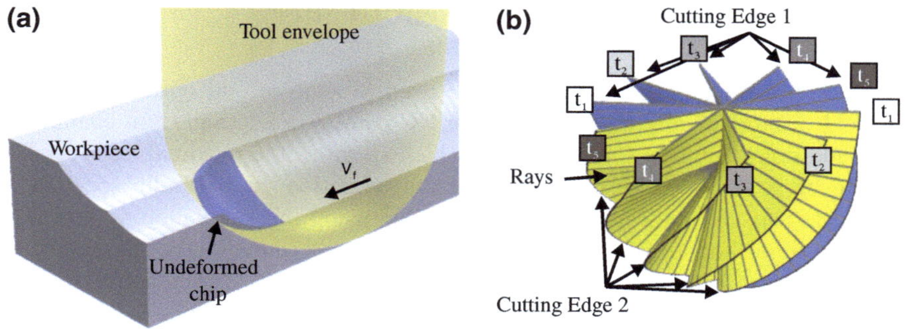

Fig. 1 **a** CSG models of tool, workpiece, and undeformed chip. **b** Scanning rays of the tool shown for two cutting edges and multiple time steps

[7]. Their parameter values, i.e., modal mass, damping, and eigenfrequency, can be determined by analyzing the measured frequency response function of the tool mounted into the spindle of the machine. The tool deflection calculated by this model at each simulated time step during one tooth feed is combined with the deflection one tooth feed earlier for the modeling of the chip shape. Thereby, the modulation of the undeformed chip thickness, which causes the regenerative effect, can be simulated. The combination of process and dynamic model is, thus, able to provide the process forces, the tool deflection, and an implicit representation of the workpiece at every simulated time step.

For an intuitive analysis of the process, an efficient visualization of the current workpiece is beneficial, which includes a color-coding of the surface location error. However, this requires an additional workpiece model since the direct rendering of a non-trivial CSG model is only possible using computationally expensive techniques, such as ray tracing [6]. The model used in the presented simulation system consists of three perpendicular dexel boards, each of which is comprised of a regularly arranged grid of depth elements (dexels) [9]. The cutting operation with this model can be performed efficiently by calculating the intersection of the dexels and the tool model. Additionally, each time a dexel is cut, the new end point of the dexel is moved along the dexel direction corresponding to the calculated tool deflection. The surface location error at this point is determined by projecting the deflection onto the surface normal so that it can be color-coded onto the displayed point (cf. Fig. 3). As will be shown in the experimental validation, this visualization technique already provides good results for 3-axis milling processes and will be extended in the future to work for arbitrary NC programs.

Modifying NC Programs to Prevent Chatter

The described simulation system can be used to analyze and optimize NC programs by automatically manipulating the NC path in order to prevent chatter. As only additional input, a stability chart, which specifies the maximal axial feed a_p leading to a stable process for a given radial feed a_e and spindle speed n, is required. This can, however, also be compiled using the simulation system. The first step in the chatter-prevention algorithm is the determination of the sections of the NC program that lead to an unstable cutting process. To accomplish this, a Fast Fourier Transform (FFT) of the recent tool displacement history is carried out during the simulation run. Chatter is assumed whenever the amplitude of a non-harmonic of the tooth engagement frequency is higher than a threshold value times the amplitude of the tooth engagement frequency itself. While identifying the unstable sections of the program, the maximum effective radial immersion $a_{e;eff}$ perpendicular to the feed direction and the maximum axial immersion ap are recorded. In this context, the effective radial feed may differ from the one preset in the CAM program, e.g., in corners of machined pockets where the increase of the effective feed is often the cause of a temporarily instable process [10] (Fig. 2a, b).

After this simulation and analysis step, the recorded maximum effective radial immersion over each instable section is used to lookup the maximum stable axial feed in the precompiled stability chart. With this information, the number of intermediate NC paths can be calculated that are required to stay below this axial feed. Additional NC segments are generated by first withdrawing the tool along the tool axis and then inserting new segments parallel to the original NC path. These parallel segments are only moved along the tool axis with a new axial feed equal to the original one divided by the new number of paths (Fig. 2c). The same segments are then used for the reverse movement and the process is repeated until the original NC path is reached. This process only directly changes the axial feed and might even decrease the maximum radial feed for some of the new segments.

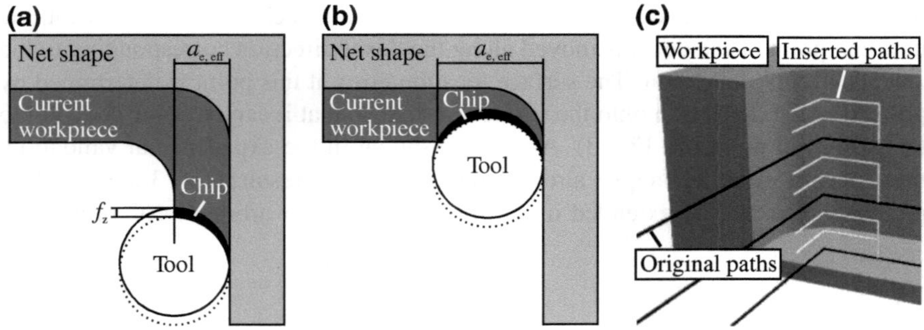

Fig. 2 **a** Effective radial feed on linear NC segments. **b** Increased effective radial feed resulting from the manufacturing of a corner. **c** Example for the insertion of new NC segments in the corner of a machined pocket

Experimental Validation

In order to validate the described simulation system and the chatter-prevention algorithm, the machining of simple pockets in aluminum (EN AW-7075) was analyzed. The NC program was generated with an axial feed $a_p = 4.5$ mm and a radial feed $a_e = 4$ mm. The pockets were then dry machined with a toroidal end mill with diameter $D = 12$ mm, corner radius $r_c = 1.5$ mm, two cutting edges, and a helix angle of 30° on a Deckel Maho DMU 50 eVolution 5-axis machining center. The process was conducted using down-milling with feed per tooth $f_z = 0.2$ mm and spindle speed $n = 15{,}800\ \mathrm{min}^{-1}$.

The used process parameter values generally result in a stable process, but chatter occurs while machining the corners of the pocket. Figure 3a and c show a corner of the workpiece generated by the simulation and the actual 3-axis milling process, respectively. Both images clearly display surface artifacts and they demonstrate that the simulation is able to predict the process instability and the direction of the surface marks. The corners produced by simulating the optimized NC program for 3-axis milling are shown in Fig. 3b and the corresponding experimental result is depicted in Fig. 3d. The same method was applied to the

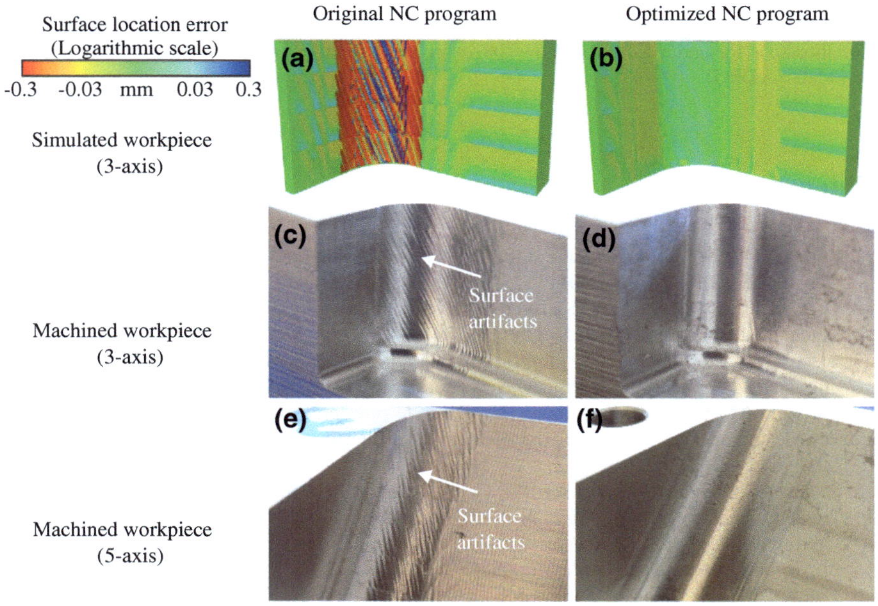

Fig. 3 Workpiece corners generated by the original (**a**, **c**, **e**) and optimized (**b**, **d**, **f**) NC programs. **a** and **b** display the simulated surface location error for the 3-axis program. **c** and **d** depict a machined corner produced by 3-axis milling. **e** and **f** show a corner produced by 5-axis milling

5-axis NC program for the corner displayed in Fig. 3e and also resulted in an optimized workpiece surface (Fig. 3f). This demonstrates that an automated optimization of NC paths with respect to chatter avoidance in critical areas is possible.

Conclusion and Outlook

In this paper, a simulation system for the five-axis milling of free-formed surfaces was presented. The described modeling techniques with an implicit representation of the undeformed chip shape and the computation of the dynamic interactions between cutting edges and workpiece allow a detailed analysis and optimization of arbitrary NC programs. In the future, the visualization model will be adapted so that surface location errors resulting from five-axis NC programs can also be displayed. However, the validation example demonstrates that the prediction of chatter is possible for different engagement conditions and can be used in combination with an NC path modification algorithm to ensure process stability. By utilizing these modeling techniques inside a CAM system, manufacturing errors could already be prevented in the early phase of NC path generation.

Acknowledgments This work is based on investigations of the research project B2 of the Transregional Collaborative Research Center (TR) 10 and of the research project B2 of the Collaborative Research Center (SFB) 708, which are kindly funded by the German Research Foundation (DFG).

References

1. Altintas, Y., Weck M.: Chatter stability of metal cutting and grinding. CIRP Ann. Manuf. Technol. **53**(2):619–642 (2004)
2. Brecher, C., Esser, M., Witt, S.: Interaction of manufacturing process and machine tool. CIRP Ann. Manuf. Technol. **58**(2):588–607 (2009)
3. Insperger, T., Stépán, G.: Semi-discretization for time-delay systems—stability and engineering applications. Springer, New York (2011) (ISBN 978-1-4614-0334-0)
4. Eksioglu, C., Kilic, Z.M., Altintas, Y.: Discrete-time prediction of chatter stability, cutting forces, and surface location errors in flexible milling systems. J. Manuf. Sci. Eng. **134**(6), 061006 (2012)
5. Zabel, A., Odendahl, S., Peuker, A.: Combining different modeling techniques to optimize the simulation of the five-axis milling process. In: Proceedings of the 5th International Conference and Exhibition on Design and Production of Machines and Dies/Molds, pp. 155–160. Kusadasi, Turkey (2009)
6. Foley, J.D., van Dam, A., Feiner, S.K., Hughes, J.F. (1992): Computer graphics—principles and practice. Addison-Wesley, Reading
7. Surmann, T., Enk, D.: Simulation of milling tool vibration trajectories along changing engagement conditions. Int. J. Mach. Tools Manuf. **47**(9), 1442–1448 (2007)

8. Kienzle, O., Victor, H.: Die Bestimmung von Kräften und Leistungen an spanenden Werkzeugen und Werkzeugmaschinen. VDI-Z **94**(11–12), 299–305 (1952)
9. Weinert, K., Stautner, M.: An efficient discrete simulation for five-axis milling of sculptured surfaces. Prod. Eng. Res. Dev. **9**(1), 47–51 (2002)
10. Ko, J.H., Shaw, K.C.: Chatter prediction algorithm in frequency domain for pocket milling. SIMTech Tech. Rep. **12**(1), 15–21 (2011)

Improved Quality of Drilled Holes in Laminated Carbon Fiber Reinforced Plastics via Laser-Preprocessing

F. Schneider, B. Kirsch, M. Gurka, T. Hermann, J. A. L'huiller and J. C. Aurich

Abstract In this paper, a new hybrid process to manufacture holes in laminated carbon fiber reinforced plastics is presented. It combines the advantages of conventional drilling (fast and easy to control) and laser drilling (high entry and exit quality of the holes). It is shown, that considerable improvements of the quality of drilled holes can be achieved via laser-preprocessing.

Keywords Cutting edge · Drilling · Fiber reinforced plastics · Laser

Introduction

Carbon fiber reinforced plastics (cfrp) combine high stiffness with low density, ideal properties for light-weight designs needed for aerospace applications and automotive, first of all for electronic vehicles [1]. Although cfrp are manufactured near net shaped, subsequent machining is necessary [2] e.g. to obtain the holes for rivet joints. These machining processes can damage the cfrp, resulting in weakening of the workpiece structure which negatively influences the operational behavior [3–5], or by thermal damage (micro-cracks, voids, delamination) due to the heat, introduced by e.g. a laser machining process [7, 8].

F. Schneider (✉) · B. Kirsch · J. C. Aurich
Institute for Manufacturing Technology and Production Systems, University of Kaiserslautern, Kaiserslautern, Rhineland-Palatinate, Germany
e-mail: schneider@cpk.uni-kl.de
URL: www.fbk-kl.de

M. Gurka
Institute for Composite Materials (IVW), University of Kaiserslautern, Kaiserslautern, Rhineland-Palatinate, Germany

T. Hermann · J. A. L'huiller
Photonik-Zentrum Kaiserslautern e.V., Kaiserslautern, Rhineland-Palatinate, Germany

B. Denkena (ed.), *New Production Technologies in Aerospace Industry*,
Lecture Notes in Production Engineering, DOI: 10.1007/978-3-319-01964-2_4,

In this paper, the quality (avoidance of delamination and burrs) of drilled holes in laminated cfrp will be examined. More specifically, the quality of mechanically drilled holes in cfrp will be compared with holes drilled with laser-preprocessing. It will be shown that improved machining quality is achieved by removing one layer at the exit of the holes via laser ablation.

Experimental Setup

The experiments were conducted on a 5-axis drilling/milling machining centre (DMU 70 eVolution). The tool used was a common two-edged uncoated drill according to DIN 6539 of 5 mm diameter with a drill-point angle σ of $\sigma = 118°$. In accordance with previous investigations [3], very sharp cutting edges were used. Using a digital fringe projection system from GFM, the rounded cutting edge radii r_β were measured to $r_\beta = 6.5$ µm. While several combinations of cutting speed v_c and feed speed f were examined, only the combination $v_c = 150$ mm/min and $f = 0.04$ mm/rev, offering the best machining quality, will be presented here.

The laser-preparation was done as a separate step using a picosecond laser system (HYPER25 Coherent Kaiserslautern GmbH) mounted on a high precision 5-axis laser micromachining system (GL.5 GFH GmbH). The laser wavelength used was 532 nm, the pulse repetition rate 200 kHz, the pulse duration about 10 ps and the spot size about 10 µm.

The material investigated in this paper was a quasi-isotropic layup [A/+45/90/−45/+45/−45/0/90/0/−45/+45/−45/90/+45/A] of 13 individual layers of a unidirectional prepreg CE1007-150-38 (0.14 mm per layer), consolidated in an autoclave for 1 h at 4 bar.

Process and Results

The quality of the drilled holes with and without laser-preprocessing was qualitatively evaluated by microscopy. The hybrid process is done in two separate steps. The first step is to preprocess either the hole entry or the hole exit with the picosecond laser pulses in the micromachining center. The second step is the drilling process on the 5-axis machining centre.

In Fig. 1 examples of the preprocessed hole entry and hole exit are depicted. In each case, the laser parameters were adjusted to exactly remove one layer of the cfrp.

In Fig. 2 the resulting quality of the hole entry with and without laser-preprocessing is shown. It can be seen that the hole entry with the removed first layer (laser-preprocessing) has fewer defects than the unprepared one. The number and size of protruding fibers were reduced.

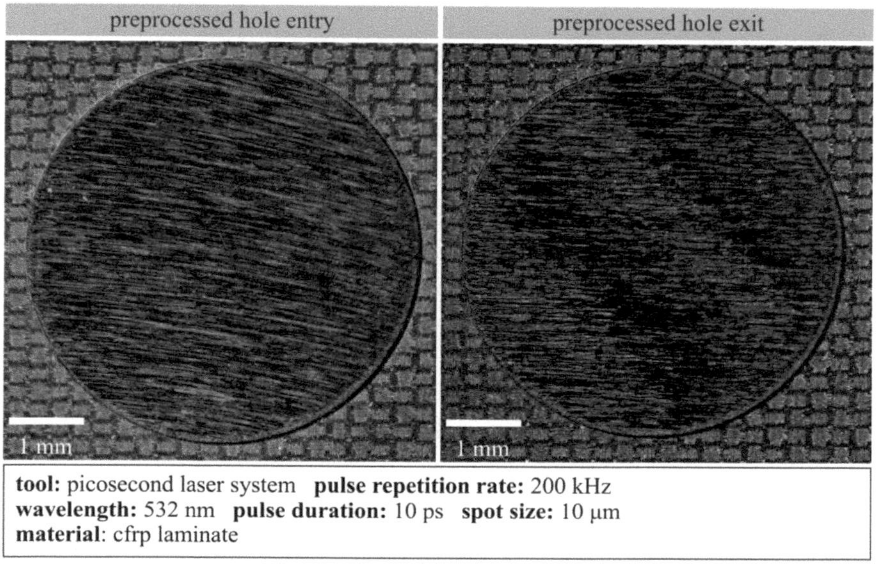

Fig. 1 Examples of laser-preprocessed hole entry and exit

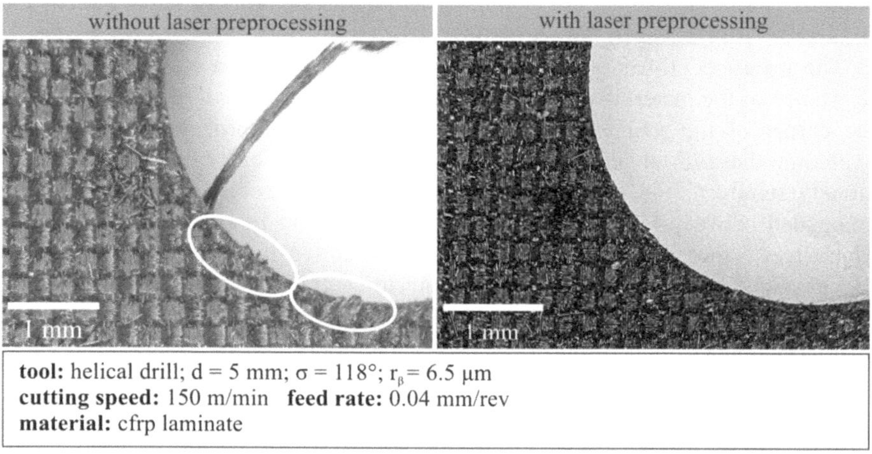

Fig. 2 Comparison of drilled hole entry with and without laser-preprocessing

This can be explained by the fact that when the first layer is removed corresponding to the diameter of the drilling tool, the force in peripheral direction is reduced. The actual drilling begins at the second layer. This layer is embedded between the first and the third layer, and hence the compound structure is more rigid. As a consequence, less peel up delamination and protruding of fibers occurs.

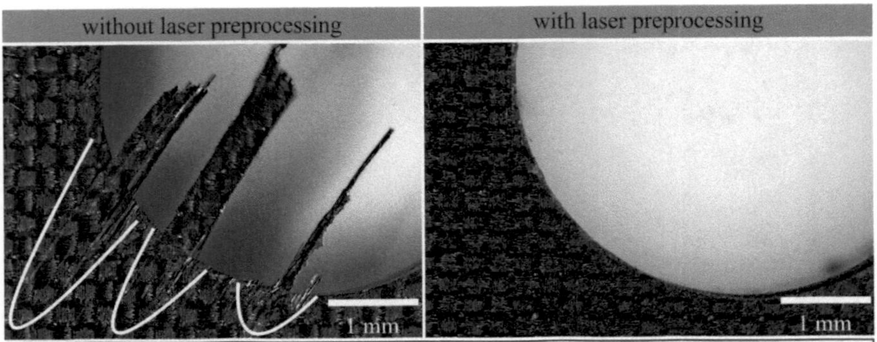

tool: helical drill; d = 5 mm; $\sigma = 118°$; $r_\beta = 6.5\ \mu m$
cutting speed: 150 m/min **feed rate**: 0.04 mm/rev
material: cfrp laminate

Fig. 3 Comparison of drilled hole exit with and without laser-preprocessing

The influence of the laser-preprocessing on the quality of the hole exit becomes even more important, as shown in Fig. 3. It can be seen that the result of the hole exit with removed last layer is much better than the one without preprocessing. The unprepared hole exit reveals very large fiber protrusion and delamination. In contrast, the laser-preprocessed hole exit verishows no unwanted effects like delamination or a frayed border area.

The protruded fibers as well as the delamination were a result of the lower resistance of the material against the drilling tool. The material is pushed away by the corner of the tool and could not be cut. Laser-preprocessing can help to overcome these disadvantages as described for the hole entry: as the last layer is already removed, the last layer to be drilled is number twelve. This layer is embedded between layer eleven and thirteen, resulting in a more rigid structure. This effect is more important for the hole exit than for the hole entry (as shown in the presented results), as there is no counterforce of the material in the hole exit. This effect was already shown in investigations on burr formation when drilling heat treated steel [6] (exit burrs are bigger than entry burrs).

Conclusion

In this paper, first results of drilled holes in laminated carbon fiber reinforced plastics with laser- preprocessing were presented. It could be shown that the quality of the hole entry and especially the hole exit could be improved considerably by removing the first and the last layer of the cfrp respectively via precise picosecond laser ablation. Advantages of this hybrid process in comparison to the manufacturing of the holes using laser machining only is a considerably reduced process time. This encouraging first results show the potential of this hybrid process. In future investigations, more laser and drilling parameters as well as different cfrp materials will be examined.

References

1. Abele, E., Kreis, M., Weigold, M.: Mit Leichtigkeit zu höherer Leistung. Trendbericht: Stand der Zerspanungstechnik im Leichtbau. In: WB Werkstatt und Betrieb 140(7–8), 66–72 (2007)
2. Aurich, J.C., Dornfeld, D., Arrazola, P.J., Franke, V., Leitz, L., Min, S.: Burrs —Analysis, Control and Removal (Keynote Paper). CIRP Ann. Manuf. Technol. **58**(2), 519–542 (2009)
3. Aurich, J.C., Zimmermann, M.: Zerspanen von Verbundwerkstoffen – Herausforderungen bei der spanenden Bearbeitung. In: Biermann, D. (Hrsg.): Spanende Fertigung – Prozesse, Innovationen, Werkstoffe. 6. Ausgabe pp. 304–312. Vulkan Verlag, Essen (2012)
4. Brinksmeier, E., Fangmann, S., Rentsch, R.: Drilling of composites and resulting surface integrity. CIRP Ann. Manuf. Technol. **60**(1), 57–60 (2011)
5. Goeke, A., Emmelmann, C.: Influence of laser cutting parameters on CFRP part quality. Phys. Proceedia **5**, 253–258 (2010)
6. Herzog, D., Jaeschke, P., Meier, O., Haferkamp, H.: Investigations on the thermal effect caused by laser cutting with respect to static strength of CFRP. Int. J. Mach. Tools Manuf. **48**, 1464–1473 (2008)
7. Hufenbach, W., Biermann, D., Seliger, G.: Serientaugliche Bearbeitung und Handhabung moderner faserverstärkter Hochleistungswerkstoffe (sefawe)-Untersuchungen zum Forschungs-und Handlungsbedarf. Institut für Leichtbau und Kunststofftechnik, Technische Universität Dresden (2008)
8. Tsao, C.C.: Thrust force and delamination of core-saw drill during drilling of carbon fiber reinforced plastics (CFRP). Int. J. Adv. Manuf. Technol. **37**, 23–28 (2008)

Flexible Production of Small Lot Sizes by Incremental Sheet Metal Forming with Two Moving Tools

Christian Magnus, Bolko Buff and Horst Meier

Abstract Incremental sheet metal forming processes have a high potential for flexible and cost efficient production of prototypes and parts in small lot sizes. Recent developments in this field bring incremental sheet metal forming one step closer to industrial application. Amongst these are improvements concerning geometrical accuracy, part complexity and surface quality. Also solutions for forming of advanced materials as magnesium or titanium alloys are being developed, which shows a high potential for the production of lightweight sheet metal parts, as used in aerospace industry.

Introduction

Current trends in the requirements for modern products show an increasing influence of individual customers on manufacturing processes. The trends of increasing part complexity, decreasing lot sizes and shorter time to market create a demand for a modernization of existing production processes. Especially in the area of sheet metal prototyping and small lot size production high costs occur for part dependent tooling, such as in stamping and for sufficient storage of dies and manufactured parts.

C. Magnus (✉) · B. Buff · H. Meier
Lehrstuhl für Produktionssysteme (LPS), Ruhr-Universität Bochum,
Universitätsstraße 150, D-44780 Bochum, Germany
e-mail: magnus@lps.rub.de

B. Buff
e-mail: buff@lps.rub.de

H. Meier
e-mail: meier@lps.rub.de

B. Denkena (ed.), *New Production Technologies in Aerospace Industry*,
Lecture Notes in Production Engineering, DOI: 10.1007/978-3-319-01964-2_5,

Fig. 1 **a** Roboforming setup, **b** setup for duplex incremental forming with local support (DPIF-L)

This paper describes developments in an incremental, robot-based sheet metal forming process ('Roboforming') for the abovementioned demand. Roboforming achieves a kinematic-based generation of a shape by using two industrial robots which are interconnected to a cooperating robot system. Compared to other incremental sheet metal forming machines, this system offers high geometrical form flexibility without the need of any part-dependent tools. Figure 1 shows the basic Roboforming setup. On each side of the fixed sheet metal one of the two robots is positioned and fitted with a universal forming tool [1]. With the help of a CAM-software (CAMWorks) and a Postprocessor (customized) the paths for both robots are derived from the CAD-Model of the desired part [2]. The whole part is formed by the movement of the robots without any need for dies. This way several different geometries can be formed with the same setup.

The industrial application of incremental sheet metal forming is still limited by certain constraints, e.g. low geometrical accuracy, part complexity, surface quality and the number of formable alloys. Within the last years many developments have contributed to an improvement in these fields.

Geometrical Accuracy

The main reasons for geometrical inaccuracies are the forming strategy, elastic deformation of the sheet and the forming machine, the positioning accuracy of the forming machine and unwanted plastic deformation of already formed areas. Especially in dieless incremental sheet metal forming some effort has to be taken in order to compensate the resulting effects.

One of the main ideas in incremental sheet metal forming with two moving tools is to have a higher flexibility in forming and to give a local support within or next to the forming zone. This way both convex and concave geometries, as shown in Fig. 2, can be formed dieless and with higher accuracy than in Single Point Incremental Forming (SPIF), which uses only one tool. Figure 1 shows a schematic setup for Duplex Incremental Forming with a Local support (DPIF-L), where both tools are in contact with the same forming zone.

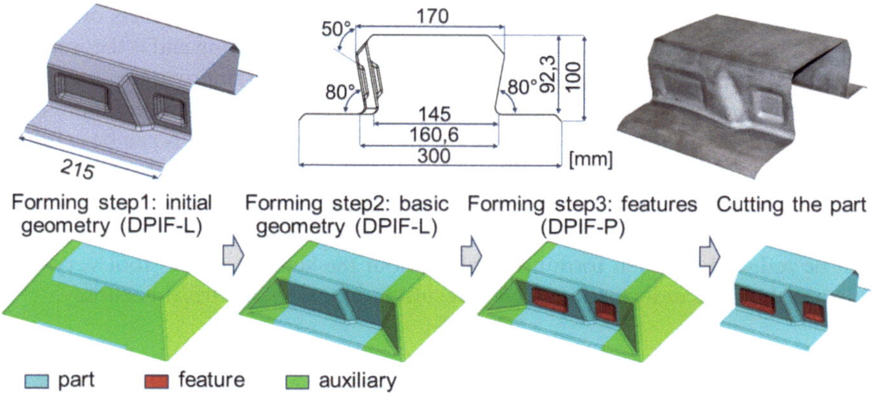

Fig. 2 Increased part complexity and flank angles by stepwise forming

As to be seen in Fig. 1b, a free area of the sheet between the part (in this case a frustum) and the clamping frame exists. Without using any kind of partial die, this area is strongly affected by unwanted plastic and elastic deformation due to the occurring forming forces. Experiments have proven that almost the same accuracy as with a partial die can be achieved by forming addendum stabilization structures which reinforce this free area [3]. Thus the partial die can be substituted. Additionally the occurring forming forces can be reduced by appropriate forming parameters or forming at elevated temperatures as it will be described later on. Also strategies for a compensation of the robots‘ low stiffness exist. These are based on measurements or predictions of the forming forces and multi body system models of the robots [4]. Today this compensation can run online for one of the robots, which guarantees for restricted workspace size almost as good accuracy as without load. Another strategy to reduce the effects of forming forces is to form the same part more than once. This can either be done with the same tool path or a path which has been corrected accordingly to the measured deviation of the part.

Especially in cold incremental sheet metal forming residual stresses exist in the formed part. In order to maintain a high geometrical accuracy even after cutting of the part, stress relief annealing can be done after unclamping of the formed part or even whilst the part is still fixed in a clamping frame.

Part Complexity

The achievable part complexity is most of all limited by a local thinning of the sheet in the formed areas and the unwanted plastic and elastic deformation of already formed areas. Although incremental sheet metal forming has increased forming limits compared to stamping [5], the maximum flank angles which can be achieved in direct forming operations are limited. Within a single forming step in SPIF maximum flank angles of about 65–70° can be formed with mild steels or aluminium. The thickness distribution is similar to that in shear spinning and is theoretically zero at 90° flank angle. If using the second tool in order to improve the stress state within the forming zone, higher formability can be achieved [6]. Furthermore it has been shown, that a high part complexity with very high flank angles can be achieved by stepwise forming operations [3, 7], as shown in Fig. 2. With the approach of stepwise forming, a more evenly thickness distribution can be achieved. By the support of the second tool the influence of the forming forces on geometric accuracy can be reduced for each forming step. Without the support of this tool very large unwanted plastic deformation occurs, resulting in low geometrical accuracy.

Surface Quality

The surface quality of the formed part in SPIF is mainly dependent on the pitch size of the tool path and the tool diameter. Both parameters influence the imprint of the tool in the surface of the part [8]. By using a second tool in DPIF-L both the inner and the outer surface quality of the part can be influenced by choosing an appropriate position of the supporting tool relatively to the forming tool and an appropriate contact force [6]. For each forming operation the tool diameter and the pitch size should be fitted to the geometric requirements of the part and the required surface quality, as the needed forming time is directly dependent on the pitch size. A small pitch size results in longer tool paths and thus in a longer forming time. For many applications (e.g. as structural components) sheet metal parts formed with a pitch size of 0.2 mm up to 0.5 mm and a tool radius of 6 mm show sufficient surface quality after short sandblasting.

Forming of Advanced Materials

In cold incremental sheet metal forming mostly mild steels and aluminium alloys are formed. Currently solutions for forming of advanced materials are being developed by several institutes. They are based on sheet metal forming at elevated temperatures, which lead to higher formability, lower forming forces, lower springback and higher accuracy [9]. For titanium [10–12], magnesium [13] and mild steels [14] the benefits of incremental sheet metal forming at elevated temperatures have been shown. Also for Complex Phase (CP) steels an improvement of formability occurs. Depending on the heating method (e.g. by laser, hot fluids, friction or resistance heating) either the whole part or a small area of the sheet is heated. Especially in the latter case a high gradient of forming characteristics, comparing those of the heated forming zone to those of the colder rest of the part, can be achieved. But, as the forming zone and the heating zone are subject to highly dynamic movement, temperature measurement and control is complex.

Summary and Outlook

Recent achievements in the fields of geometrical accuracy, the number of formable alloys and part complexity show a great improvement of the applicability of this process. The ongoing research in these fields is also focused on advanced materials which have high potential for lightweight applications. Future work of the authors will e.g. focus on parts of larger scale and higher complexity, geometrical accuracy and forming of different materials at elevated temperatures.

Acknowledgments This research is funded by the German Research Foundation (DFG). The authors are responsible for the contents of this publication.

References

1. Meier, H., Buff, B., Smukala, V.: Robot-based incremental sheet metal forming—increasing the part accuracy in an automated, industrial forming cell. KEM **410–411**, 159–166 (2009)
2. Meier, H., Zhu, J., Buff, B., Laurischkat, R.: CAx process chain for two robots based incremental sheet metal forming. Procedia CIRP **3**, 37–42 (2012)
3. Buff, B., Magnus, C., Zhu, J., Meier, H.: Robot-based incremental sheet metal forming—increasing the geometrical complexity and accuracy. KEM **549**, 149–155 (2013)
4. Meier, H., Laurischkat, R., Bertsch, C., Reese, S.: Prediction of path deviation in robot based incremental sheet metal forming by means of an integrated finite element—multi body system model. KEM **410–411**, 365–372 (2009)
5. Allwood, J.M., Shouler, D.R., Tekkaya, A.: The increased forming limits of incremental sheet forming processes. KEM **344**, 621–628 (2007)
6. Meier, H., Magnus, C., Smukala, V.: Impact of superimposed pressure on dieless incremental sheet metal forming with two moving tools. CIRP Ann. Manuf. Technol. **60**, 327–330 (2011)
7. Tanaka, S., Hayakawa, K., Nakamura, T.: Incremental sheet forming with direction control of path planes. In: Proceedings of the 10th ICTP, pp. 503–507 (2011)
8. Hirt, G., Junk, S., Witulski, N.: Surface quality, geometric precision and sheet thinning in incremental sheet forming. In: Proceedings of Materials Week, pp. 1–8 (2001)
9. Neugebauer, R., Altan, T., Geiger, M., Kleiner, M., Sterzing, A.: Sheet metal forming at elevated temperatures. CIRP Ann. Manuf. Technol. **55**, 793–816 (2006)
10. Duflou, J., Callebaut, B., Verbert, J., Baerdemaeker, H.: Laser assisted incremental forming: formability and accuracy improvement. CIRP Ann. Manuf. Technol. **56**(1), 273–276 (2007)
11. Fan, G., Sun, F., Meng, X., Gao, L., Tong, G.: Electric hot incremental forming of Ti-6Al-4 V titanium sheet. Int. J. Adv. Manuf. Technol. (Springer London) **49**, 941–947 (2010)
12. Ambrogio, G., Filice, L., Gagliardi, F.: Formability of titanium alloys in incremental sheet forming process with local material heating. In: Proceedings of the 10th ICTP, pp. 536–540 (2011)
13. Galdos, L., Saenz de Argandona, E., Ulacia, I., Arruebarrena, G.: Warm incremental forming of magnesium alloys using hot fluid as heating media. KEM **504–506**, 815–820 (2012)
14. Meier, H., Magnus, C.: Incremental sheet metal forming with direct resistance heating using two moving tools. KEM **554–557**, 1362–1367 (2013)

Dedicated Machine Tool Development for Blisk Milling

B. Bringmann, R. Bacon and B. Güntert

For modern jet engines, more and more turbine blade stages are made as one integral part in order to reduce weight.These parts—called Blisks ("Blade integrated disks") or IBRs ("Integrally bladed rotors")—are machined predominantly from Titanium or Nickel-based super alloys (see e.g. [1]). From a machining point of view, they are extremely challenging. The materials used belong to the hardest to cut, the geometries are difficult free-forms and the finished parts are flimsy. Due to the extreme requirements on these parts the machining time and the cost per workpiece are very high today. Despite the challenges of these workpieces, usually standard 5-axis machines are used for machining. With dedicated machine tools the productivity could be increased tremendously. Here the process of machine tool development especially for Blisk milling is presented, using state of the art scientific methods.

Introduction

Blisk machining combines very high requirements in machining dynamics, surface finish and—due to the long cycle times (between 20 and >200 h)—thermal stability. With a machine tool designed for the purpose, the productivity and quality of such a milling process can be improved dramatically. This approach to find the best possible design for this application is described here.

B. Bringmann (✉) · R. Bacon · B. Güntert
Starrag AG, Seebleichestr. 61, 9404 Rorschacherberg, Switzerland
e-mail: bernhard.bringmann@starrag.com

B. Denkena (ed.), *New Production Technologies in Aerospace Industry*,
Lecture Notes in Production Engineering, DOI: 10.1007/978-3-319-01964-2_6,

Analysis of the Machining Process

This process starts by identifying the desired optimal machining process in a first step. With this knowledge the requirements for the machine can be quantified.

In this first step the most efficient milling strategies are evaluated for typical Blisks. Typical approaches today are e.g. plunge milling, the use of big disk cutters for slotting or the use of ceramic tools with very high cutting speeds for Nickel-based super alloys.

The cutting parameters achievable with the tools today (and anticipated future cutting parameters) are taken as a basis. This information can then be processed in different ways. For the different operations spindles speed, power and cutting forces can be computed. This computation is very straightforward. In general the well-known basic models can be used. The machine dynamics and stability to be able to make fully use of these performance parameters, i.e. to machine productively, must be determined with more effort.

Selection of the Kinematic Concept of the Machine

In general, the most appropriate kinematic build-up for Blisk machining process should be selected. In addition to the usual design criteria like accessibility, ergonomics, floor space, etc. there are characteristics that are not very straightforward to evaluate, especially the machine dynamics for the application to keep finishing times as low as possible and the process stability to be able to us the optimal roughing strategies.

Determination of Machine Dynamics with FEM and Virtual NC Kernel

For this application machine dynamics is evaluated by the actual time that a machine will need to do a typical Blisk finishing operation. This time is determined by many factors.

The kinematic build-up determines how the different axes of the machine must move. Typically the compensation motions of the linear axes to follow the movements of the rotary axes are depending on this build-up. When e.g. a trunnion is turning, two linear axes must move just to keep the same relative position between tool and workpiece.

In addition, this finishing time is of course depending on the axis dynamics parameters. The maximum axis velocity can be determined easily from the mechanical parameters. The achievable acceleration and jerk values when taking into account certain requirements for a dynamic path accuracy and effects like overshoot, cross-talk or in-talk (see e.g. [2, 3, 4]) are more difficult to determine.

With modern simulation programs, such as a "Virtual NC Kernel" [5], very powerful tools are available to determine the effect of both kinematic build-up and axis dynamics on total cycle time. With this different machine concepts can be compared and very accurate machining times can be determined in a very early design stage. It can be computed which parameters (like accelerations) have which effect on cycle time and therefore the ideal configuration can be identified. With this different kinematic concepts and variations can be compared.

This VNCK uses the same inputs as a real control and generates the same outputs. One drawback is that—while axis limitations like filters or max. jerk values can easily be put in—it cannot be determined what values have to be put in because these depend on the mechanical behaviour. Realistic values can be determined from FEM. This approach is shown in Fig. 1.

The mechanical constraints can be anticipated by FEM simulations. Especially important here is the behaviour of the single axis against quasi-static forces (how much will the axis tilt due to acceleration forces, thus determining cross-talk and in-talk) and the Eigen frequencies when excited through the axis drive (these Eigen frequencies limit the bandwidth of the control loop and subsequently limit the achievable gains, filter and jerk settings).

With the results from these FEM simulations (see e.g. Fig. 2) the parameters for VNCK can be determined. With this the operation can be simulated, the time can be determined and limitations preventing shorter times can be simulated.

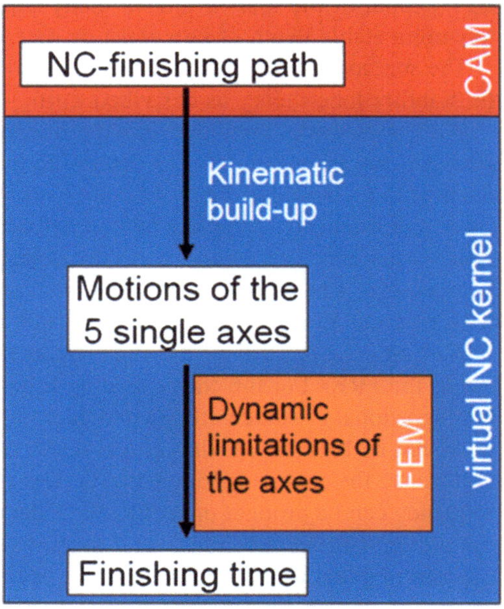

Fig. 1 Simulation approach for determining machine dynamics using VNCK and FEM

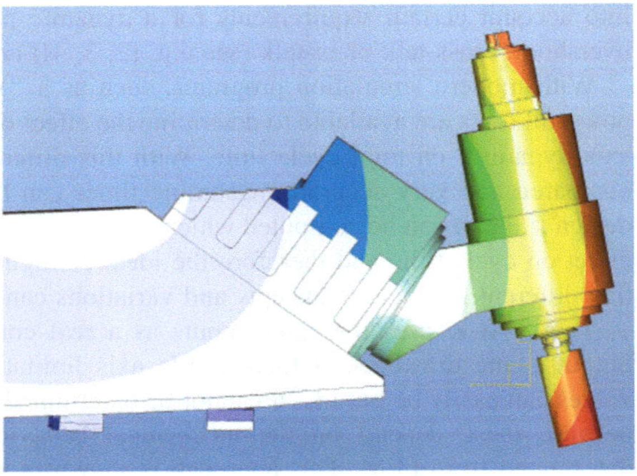

Fig. 2 Example of FEM result: Eigen mode of an assembly

This first result can then be used to overcome limitations, e.g. by changing the kinematic build-up, by systematic mechanical improvements where beneficial, by improving the control optimization.

Determination of Process Stability with FEM and CutPro

The other factor not easy to determine is the process stability of the system. For the Blisk machine development it is crucial that the most efficient strategies and the potential of the tools available today and tomorrow can be used. For this the maximum stable depth of cuts must be determined.

The method of choice today are software packages such as "CutPro", using the stability lobe theory as described in [6].

While this works excellent for experimental analysis when a certain tool in the spindle is tap tested in order to determine the frequency response function, for a machine to be built this data is of course not yet available. A good alternative is to determine frequency response functions from FEM. To get good results, there is some experience necessary for the assumptions of stiffnesses and damping parameters of single components. With that prior knowledge, very good agreement between simulation and experimental result are possible.

With the results, the most critical frequency can be determined (smallest real part in the FRF). For the corresponding Eigen mode, the critical depth of cut will be proportional to static stiffness and damping ratio. If the Eigen frequency of this mode can be brought up, the critical depth of cut a_{lim} will change approximately with square the change in frequency. So if e.g. by design changes the critical Eigen frequency can be brought up to 120 % of the initial state, a first estimate would be that the critical depth of cut would increase to 144 % from the initial state. This assumes in general unchanged damping behaviour.

Selected Machine Tool Concept

Initial simulations with trunnion concepts following the methods described above showed that with such concepts, there were two constraints limiting the dynamic of the system.

The trunnion axis carrying the second rotary axis, the pallet, the fixture and the workpiece was one limiting factor because of the limited acceleration possible due to the high inertia. The other one were the dynamics of the linear axes because, with a trunnion machine, rather big compensation movements are necessary to follow the trunnion.

So to really come to significantly shorter cycle times, both the rotary axis had to be accelerated and the necessary compensation motion reduced.

The solution was the use of a tilted B-axis as known in machining of single turbine blades as seen in Fig. 3. The total machine concept is shown in Fig. 4.

This concept combines two advantages:

Instead of tilting a trunnion with the very high inertia, the B-axis only has to rotate the main spindle, drastically reducing the inertia. With that, much higher axis accelerations are possible. Due to the stiff build-up with very high Eigen frequencies, also very high jerk limits become possible.

Due to the tilted design, the tool center point (TCP) approximately lies in the geometric B-axis. That means that when rotating the tool the tool center point almost does not move (depending on tool length). Therefore no compensation motions to follow this rotation are necessary.

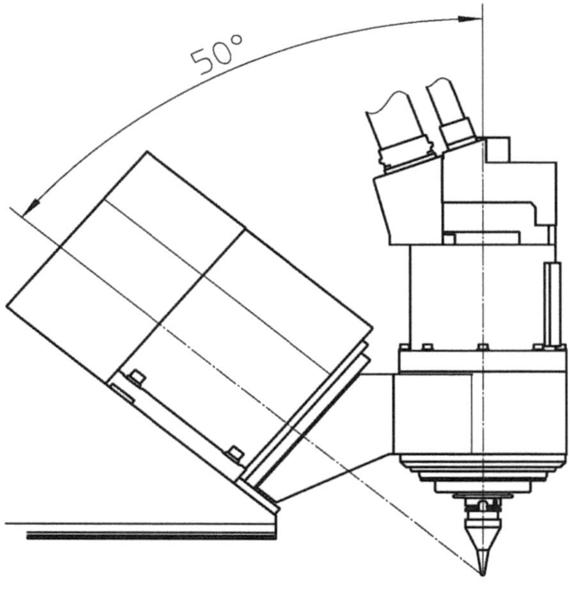

Fig. 3 Side view on spindle and B-axis with a 50° angle in between to allow for tilt motions around the tool center point (TCP)

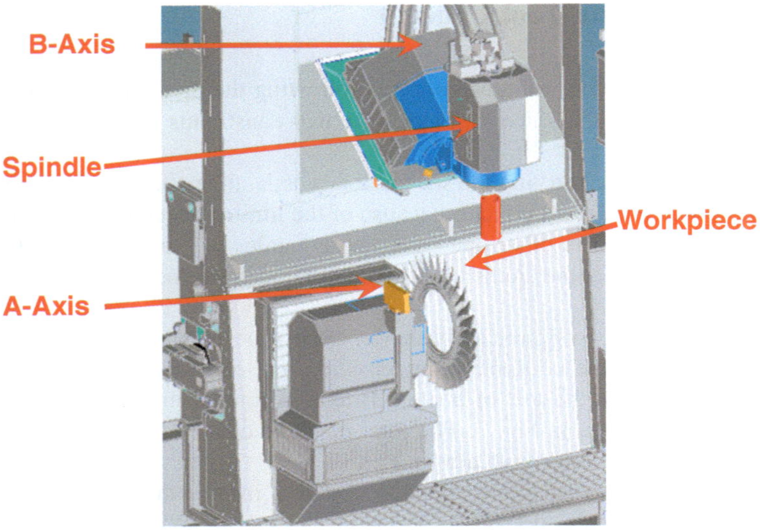

Fig. 4 Machine concept with tilted B-axis

In comparison to trunnions, the simulations have shown that the compensation motions when swivelling are usually reduced by more than 90 % and the time required for such swivelling motions is reduced by about 80 %.

Conclusion

To design a machine, new development tools do exist that allow for very accurate determination of the machine behaviour at a very early design stage. With this methods, it can already be determined if e.g. roughing cuts can be performed as foreseen. Already control parameters can be estimated and cycle times can be anticipated in great details. Bottlenecks limiting the performance can be identified and eliminated. Due to the concentration on such a special application as Blisk milling, breakthroughs in performance can be achieved.

References

1. Blair, L.W., Tapparo, D.J.: Axial-centrifugal compressor program. (Department of Defense, Department of the Army, Army Materiel Development and Readiness Command, Army Aviation Research and Development Command), Army Air Mobility Research and Development Laboratory, Eustis Directorate, Houston (1975)

2. Nguyen, M., Weikert, S., Wegener, K.: Evaluation method of acceleration correlated position errors in machine tools. Institute of Machine Tools and Manufacturing, Swiss Federal Institute of Technology Zurich, Zurich (2012)
3. Steinlin, M., Weikert, S., Wegener, K.: Open loop inertial crosstalk compensation based on measurement data. In: 25th Annual Meeting of the American Society for Precision Engineering, The American Society for Precision Engineering, Raleigh, USA (2010)
4. Bringmann, B., Maglie, P.: A method for direct evaluation of the dynamic 3D path accuracy of NC machine tools. Ann. CIRP **58**(1), 343–346 (2009)
5. Siemens AG: SINUMERIK 840D VNCK Reference (FBVNCK), 9th edn. Siemens, Munich (2007)
6. Altintas, Y., Budak, E.: Analytical prediction of stability lobes in milling. Ann. CIRP **44**(1), 357–362 (1995)

Surface Characterization of Components Subjected to Deep Rolling for Cyclic Loading Applications

A. M. Abrão, B. Breidenstein, T. Mörke and B. Denkena

Abstract Deep rolling is mainly employed to induce compressive residual stresses and cold work straining on cyclically loaded parts. Additionally, surface roughness can be improved and, as a result, the fatigue strength of the component is increased. The principal aim of this work is to investigate the influence of deep rolling parameters (rolling pressure and number of passes) on the surface quality and residual stresses induced on AISI 1060 high carbon steel specimens. The results indicated that the surface roughness parameters decrease drastically after deep rolling and lowest roughness values are obtained applying a pressure of 100 bar and 3 rolling passes. As far as the residual stress is concerned, the tensile stress recorded on the surface of the specimens after turning shifted to compressive stress after deep rolling. In contrast to the residual stress values, which did not present a straightforward relationship with deep rolling parameters, the full width at half maximum values indicated that more accentuated work hardening is obtained when rolling pressure is elevated to 200 MPa.

Introduction

Deep rolling is a non-cutting production method which aim is to introduce work hardening and compressive residual stresses into the surface and subsurface layers in order to increase fatigue strength [1]. Furthermore, the surface roughness is improved. For this purpose, the rolling tool is pressed against the part over a

A. M. Abrão (✉)
Department of Mechanical of Engineering, Universidade Federal de Minas Gerais,
Av. Antônio Carlos 6627, Pampulha, Belo Horizonte, MG 31270-901, Brazil
e-mail: abrao@ufmg.br

B. Breidenstein · T. Mörke · B. Denkena
Institute of Production Engineering and Machine Tools, Leibniz Universität Hannover,
An der Universität 2, 30823 Garbsen, Germany

B. Denkena (ed.), *New Production Technologies in Aerospace Industry*,
Lecture Notes in Production Engineering, DOI: 10.1007/978-3-319-01964-2_7,

determined number of passes to cause plastic strain and depending on the selected parameters distinct levels of residual stress and surface finish can be obtained.

An increase in the yield stress leading to an elevation of approximately 20 % in the fatigue strength of steels specimens subjected to deep rolling is reported by [2]. Similarly, Pengfei et al. [3] state that the ultimate tensile strength of a low carbon steel increases by 18.6 % after deep rolling. These findings suggest an appreciable elevation in the specific strength of deep rolled components, thus making this process attractive for applications where low weight is relevant.

In general, the maximum hardness value resulting from plastic deformation after deep rolling is not found on the work surface, but beneath it where the Hertzian stress reaches its maximum during the operation. Nevertheless, an additional increase in yield strength and hardness can be obtained by reducing grain size (Hall–Petch effect), as reported by [3].

Applied pressure, number of passes, ball diameter, rolling speed and feed rate are the principal factors affecting the residual stress profile and surface finish. The influence of ball diameter and applied pressure when deep rolling a cold work tool steel with average hardness of 400 HV was reported by [4]. Lower surface roughness values were obtained using the ball with largest diameter (Ø13 mm) and highest pressure (400 bar).

Medium carbon steel parts subjected to deep rolling with a tungsten carbide ball (Ø5 mm) presented a substantial decrease in roughness with the number of passes [5]: arithmetic average height Ra equal to 4.01 μm after turning, 0.46 μm after the first pass, 0.10 after the second pass and 0.09 μm after the third pass (maximum peak to valley height Rt = 14.4 μm, then 2.76 μm followed by 0.78 μm and finally 0.60 μm, respectively).

According to [6] there is a great level of interaction between the deep rolling parameters and the generated surface texture, therefore, in order to reduce the roughness of deep rolled medium carbon steel (187 HV) it is recommended to employ high values for rolling speed, pressure and number of passes and small ball diameters.

This work is focused on the influence of deep rolling parameters (pressure and number of passes) on the surface finish and residual stress induced on turned AISI 1060 high carbon steel specimens.

Experimental Procedure

Bars of AISI 1060 high carbon steel (Ø20.3 × 135 mm) were used as work material. The experimental method is depicted as follows: firstly, the specimens were rough machined with geometry in accordance with [7] but with an oversize of 2 mm in diameter. The samples were then heat treated to the annealed state (heating to a maximum temperature of 660 °C followed by furnace cooling), thus resulting in an average hardness of 260 HV0.5. After that, finish turning was carried out in two passes employing the following cutting conditions: pre-finishing at cutting speed (v_c) of 200 m/min, feed rate (f) of 0.3 mm/rev and depth of cut (a_p) of 0.25 mm and finishing at $v_c = 100$ m/min, f = 0.1 mm/rev and maximum $a_p = 0.15$ mm. Coated tungsten carbide inserts grade 4,215 and geometry DNMG 110408-PM were mounted on a tool holder code DDNNN 2020 K11 and used as cutting tools (inserts and holder supplied by Sandvik Coromant). Finally, the specimens were subjected to deep rolling under four distinct conditions: rolling pressures (p) of 100 and 200 bar and number of passes (n) of 1 and 3 and using constant rolling speed and feed rate values ($v_r = 100$ m/min and $f_r = 0.07$ mm/rev, respectively). Four samples were produced under each condition, thus resulting in a total of 20 runs (including 4 samples which were not deep rolled and used as control).

Turning and deep rolling were performed in a high stiffness Gildemeister CTX 520 CNC lathe (145 kW power and 10,000 rpm maximum rotational speed). An Ecoroll tool connected to a HGP 1.0 hydraulic pump (maximum pressure of 400 bar) was used for deep rolling. This device possesses 3 tungsten carbide balls (Ø6 mm) equally spaced which simultaneously deep roll the work surface when the required pressure is applied. Therefore, n = 1 and 3 mean that the surface was rolled 3 and 9 times, respectively.

After the manufacturing of the samples, the assessment of the generated surfaces was conducted in accordance with [8] standard using a Nanofocus μ surf confocal microscope (sampling length of 0.8 mm) and software MountainsMap 5.1. A Zeiss EVO 60 scanning electron microscope was employed for visual inspection of the specimens' surface. Surface axial residual stress and the corresponding full width at half maximum values were measured using a GE Inspection Technologies X-ray diffraction system model XRD 3003 TT with 2 mm diameter point collimator. The $\sin^2 \Psi$ method was employed using CrKα radiation on 211 planes of the ferrite phase and varying Ψ from $-45°$ to $45°$.

Results and Discussion

The influence of distinct values of pressure (p = 100 and 200 bar) and number of passes (n = 1 and 3) on selected surface roughness parameters (arithmetic average height Ra, root mean square roughness Rq, maximum peak to valley height in the sampling length Rz, maximum peak to valley height in the assessment length Rt, skewness Rsk, kurtosis Rku, reduced valley depth Rvk, core roughness depth Rk and reduced peak height Rpk) is given in Fig. 1. The findings concerned with the amplitude parameters Ra, Rq, Rz and Rt (Fig. 1a and b) indicate a remarkable reduction in roughness after deep rolling, irrespectively of the parameters used, however, lowest values are obtained applying p = 100 bar and n = 3 and the further elevation of rolling pressure to p = 200 bar impairs the surface finish probably due to plastic flow of the work material.

As far as the hybrid parameters are concerned (Fig. 1c), Rsk is sensitive to occasional high peaks or deep valleys [9] and a null skew would be expected for a

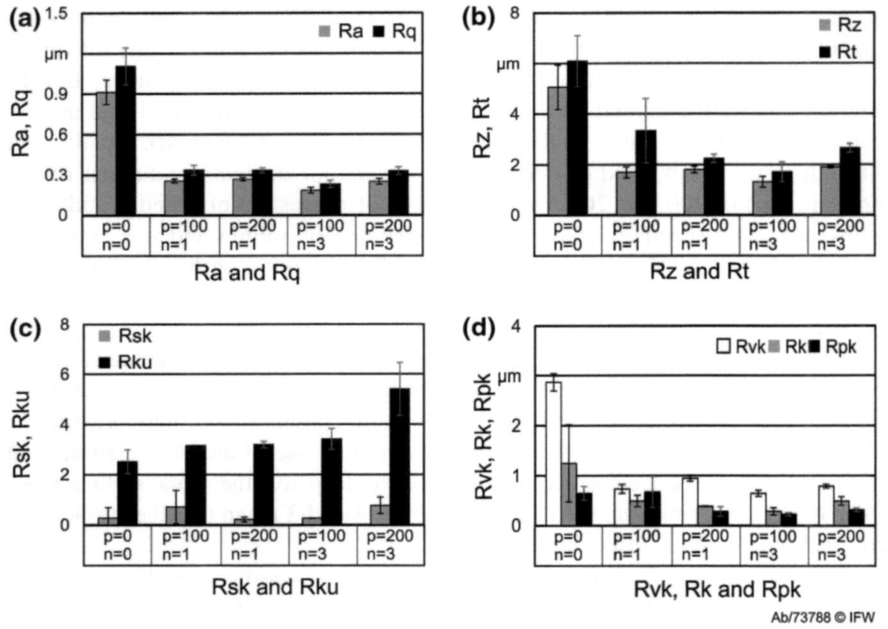

Fig. 1 Effect of deep rolling pressure and number of passes on surface roughness: **a** Ra and Rq, **b** Rz and Rt, **c** Rsk and Rku and **d** Rvk, Rk and Rpk

surface with symmetrical height distribution (as many peaks as valleys), whereas profiles with filled valleys present Rsk > 0, which seems to be the case. Furthermore, Rku gives an indication of the sharpness of the curve and a kurtosis value of approximately Rku = 3 suggests a Gaussian distribution of the profile amplitude, while Rku < 3 (turned surface, p = 0 and n = 0) denotes a surface with relatively few high peaks and low valleys and Rku > 3 indicates a surface with many relatively high peaks and low valleys [9]. Finally, the results obtained from the material ratio curve indicate a considerable reduction in the core roughness depth (Rk) and in the reduced peak height (Rpk) after deep rolling due to plasticization, see Fig. 1d.

Photographs of the surfaces processed under distinct deep rolling conditions and using different magnifications are presented in Fig. 2, where the feed marks left by the cutting tool can be clearly seen in Fig. 2a. After deep rolling using p = 100 bar and n = 1 (Fig. 2b and c), feed marks are no longer visible as plasticization takes place. An increase in rolling pressure to p = 200 bar (Fig. 2d and e) leads to accentuated plasticization. Finally, when the number of passes is increased to n = 3, see Fig. 2f, surface plastic deformation becomes evident.

Surface residual stress and full width at half maximum results are presented in Fig. 3a and b, respectively. It can be seen that the tensile residual stress induced by turning (average of 207 MPa) shift to compressive values after deep rolling (below

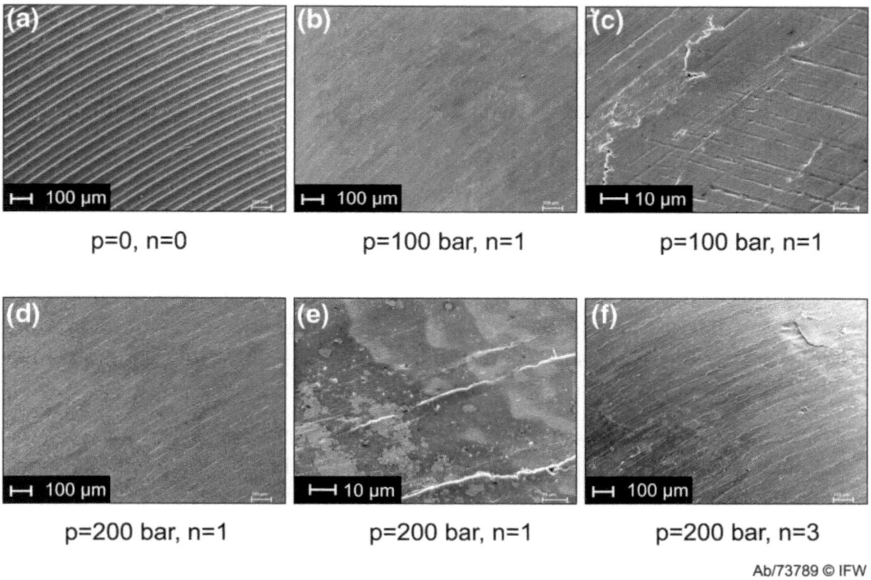

Fig. 2 Effect of deep rolling pressure and number of passes on surface texture: **a** $p = 0$, $n = 0$ (turned); **b** and **c** $p = 100$ bar, $n = 1$; **d** and **e** $p = 200$ bar, $n = 1$ and **f** $p = 200$ bar, $n = 3$

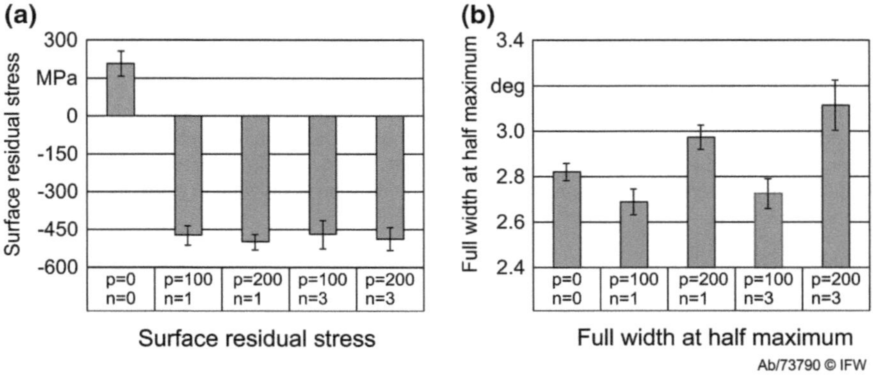

Fig. 3 Effect of deep rolling pressure and number of passes on: **a** surface residual stress and **b** surface full width at half maximum

−450 MPa), however, no relationship can be drawn with regard to the influence of the rolling parameters, probably owing to the fact that according to the Hertz contact pressure theory the maximum equivalent stress is found beneath the surface. Nevertheless, the full width at half maximum values suggest that distinct levels of work hardening are induced depending on the rolling pressure, i.e., higher values were obtained applying a rolling pressure of p = 200 bar. This could be an indication that higher dislocation density is obtained when higher rolling pressure values are employed. Interestingly, no distinction can be made between the full width at half maximum values with regard to the nature of the residual stress (tensile or compressive).

Concluding Remarks

After subjecting annealed AISI 1060 high carbon steel specimens to deep rolling under distinct rolling pressures and number of passes the following conclusions can be drawn:

- A substantial improvement on the surface roughness amplitude parameters is obtained after deep rolling, with lowest values being obtained applying a pressure of p = 100 bar and number of rolling passes n = 3.
- As rolling pressure and number of passes are increased, more intense plastic deformation and material flow takes place on the work surface, thus impairing the quality of the surface.
- The tensile residual stress induced on the surface by turning shifts to compressive stress after deep rolling, however, the influence of the rolling parameters is not evident. In contrast, higher full width at half maximum values are recorded after deep rolling at higher pressure (p = 200 bar).

Acknowledgments The authors would like to thank the Institute of Materials Science of the Leibniz Universität Hannover for the heat treatment of the specimens and both Ecoroll AG Werkzeugtechnik (Celle, Germany) and Sandvik Tooling Deutschland GmbH for the provision, respectively, of the deep rolling and cutting tools. A. M. Abrão is grateful to the CAPES Foundation, Ministry of Education of Brazil, for funding his scholarship (grant no. 10118128).

References

1. Schultze, V.: Modern mechanical surface treatment, 368 pp. Wiley-VHC Verlag GmbH & Co. KGaA, Weinheim (2006)
2. Berstein, G., Fuchsbauer, B.: Festwalzen und Schwingfestigkeit. Zeitschrift für Werkstofftechnik **13**, 103–109 (1982)
3. Chui, P., Sun, K., Sun, C., Chengge, Wu, Wang, H., Zhao, Y.: Effect of surface nanocrystallization induced by fast multiple rotation rolling on mechanical properties of a low carbon steel. Mater. Des. **35**, 754–759 (2012)
4. Meyer, D., Brinksmeier, E., Hoffmann, F.: Surface hardening by cryogenic deep rolling. Procedia Eng. **19**, 258–263 (2011)
5. Morimoto, T.: Work hardening and tool surface damage in burnishing. Wear **127**, 149–159 (1988)

6. Seemikeri, C.Y., Brahmankar, P.K., Mahagaonkar, S.B.: Investigations on surface integrity of AISI 1045 using LPB tool. Tribol. Int. **41**, 724–734 (2008)
7. DIN 50125 Test of metallic materials—Tensile test pieces (in German), 14 pp (2009)
8. DIN EN ISO 4287 geometrical product specification (GPS)—Surface texture: profile method—terms, definitions and surface texture parameters (German version EN ISO 4287:1998+AC:2008+A1:2009), 14 pp (2009)
9. Gadelmawla, E.S., Koura, M.M., Maksoud, T.M.A., Elewa, I.M., Soliman, H.H.: Roughness parameters. J. Mater. Process. Technol. **123**, 133–145 (2002)

Small-Scaled Modular Design for Aircraft Wings

L. Overmeyer and A. Bentlage

Introduction

The product design already determines 70 % of the future product costs [1]. Therefore, the structural design offers a promising approach for reducing the lifecycle costs of an airplane. An essential design parameter is the decision between a monolithic or modular construction. The current design of aircraft wings can be described as largely monolithic; the basic structure consists of a small number of undivided components. For example the top cover of the Airbus A350 wing consists of a single component with a length of 32 m [2]. For the production of these large scaled components production machines, e.g. autoclaves, with very huge dimensions are needed. However other aircraft parts such as the fuselage are not one-piece components. The fuselage is composed of several fuselage modules, which are pre-equipped and then riveted together. In this paper, a small-scaled modular design for airplane wings is presented and the technical feasibility is discussed. Moreover, we identified modularization factors which significantly influence the decision between a monolithic or modular construction.

Literature Review

First modular constructions of large scaled products already exist in the industry. In addition to the example of the aircraft fuselage also container ships [3] and some rotor blades of wind turbines [4] are prefabricated in smaller modules and

L. Overmeyer (✉) · A. Bentlage
IPH—Institute of Integrated Production Hanover gGmbH (non-profit limited company), Hanover, Germany
e-mail: bentlage@iph-hannover.de

B. Denkena (ed.), *New Production Technologies in Aerospace Industry*,
Lecture Notes in Production Engineering, DOI: 10.1007/978-3-319-01964-2_8,

then assembled. Current scientific approaches to modularization only deal with conventional products and do not consider large-scale products such as aircraft or components of aircrafts, e.g. [5, 6]. Therefore, the term modularity is often 'only' used for the specific grouping of components to subsystems as independent as possible. There is no methodology available to decide whether a modular design is useful and which module size leads to the lowest costs.

In order to allow a differentiation between modularization approaches for conventional products and a new modularization approach for large-scaled products the term small-scaled modular design is introduced. The distinctiveness of this approach is the possibility to separate large components. A methodical approach to determine the ideal module size for large scaled products was developed. The approach is divided into the steps: design of a small-scaled modular product, check of the technical feasibility, analysis of the economic viability and development of a tool to determine the ideal module size. In the scope of this paper, the emphasis is laid on the design of a modular wing and the technical feasibility.

Design of a Small-Scaled Modular Wing

In the first step a small-scaled modular product has to be developed. We selected one of the most modern aircraft wings of a long-range, wide-body jet airline as a reference product. While maintaining the basic wing shape, the geometric dimensions and the material, different design concepts were developed. To support the systematic development a morphological box was used (see Fig. 1).

The chosen set of parameters of the morphological box does not only form the basis for the design concept but has also a significant influence on the manufacturing process. For example, the parameter 'basic shape' of the wing box determines beside design details especially in which steps the assembly of the wing box is done, which has a significant influence on the accessibility during the subsequent wing equipping.

Figure 1 shows in an exemplary manner four rough design concepts of a small-scaled modular wing. In order to identify the best design concept, all concepts were evaluated on the basis of the following criteria: manufacturability of the components, easy assembly with a high proportion of preassembly, the use of state of the art technologies, the estimated weight and the comparability to the reference wing. Using this evaluation method variant 4 was chosen, especially because this variant shows a high level of comparability to the reference wing and is based upon currently available production technologies.

The wing box construction of this variant is comparable to current airplane wings. Each module consists of a front spar, a rear spar, ribs and two covers. Because of this construction every module can be produced independently. Moreover, the tightness and geometric shape can be tested separately. The connection of the modules through riveting overlapping elements allows in addition a compensation of production tolerances. The developed concept also allows a

	value	parameter						
wingbox	basic shape	closed box (conventional) ■	U-shaped box with ▲ flat cover	● two U-shaped boxes	box with open spars ◆	multi spar design with flat cover		
wingbox	connection	riveting ■ overlapping elements	riveting ▲ doubler sheets	bonding ●	laminating	screwing	clinching	welding ◆
TE/ LE	connection LE	mounting brackets ■	▲ stay bolts	hooking into groove	● guide rail	riveting ◆ to the wingbox		
media supply	cable position	inside of front-/rear spar	▲ outside of front-/rear spar	■ between the ribs ● of the LE/TE	bottom cover of ◆ wingbox			
media supply	connection	plug-in connection■	▲no disconnection	soldering ●	screw connection ◆			

legend: TE – trailing edge; LE – leading edge; ▲ - variant 1; ◆ - variant 2; ● - variant 3; ■ - variant 4

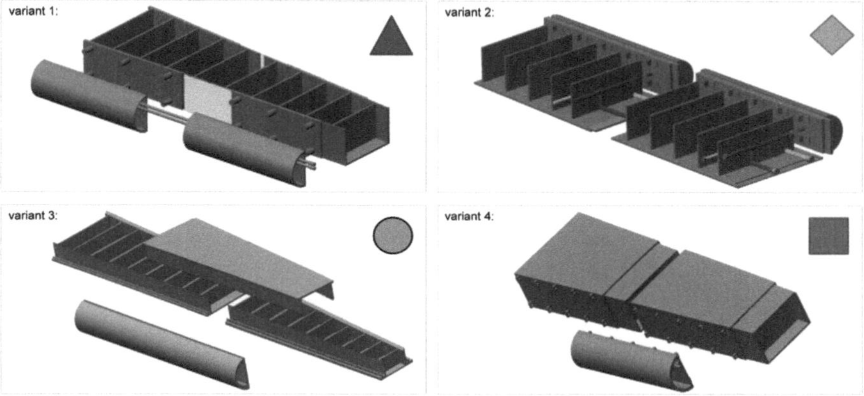

Fig. 1 Morphological box to design a modular wing

separation of the wing box perpendicular to the span width of a wing. For separations of this kind doubler sheets are used to connect the parts of the wing cover. This ensures that there is adequate assembly space available for a connection of separated ribs. The spars of the wing box modules are already equipped with mounting devices for the leading and trailing edge, e.g. by the use co-bonding or co-curing in the production of the spars. This allows a preassembly of the leading and trailing edge with very good accessibility (see Fig. 2).

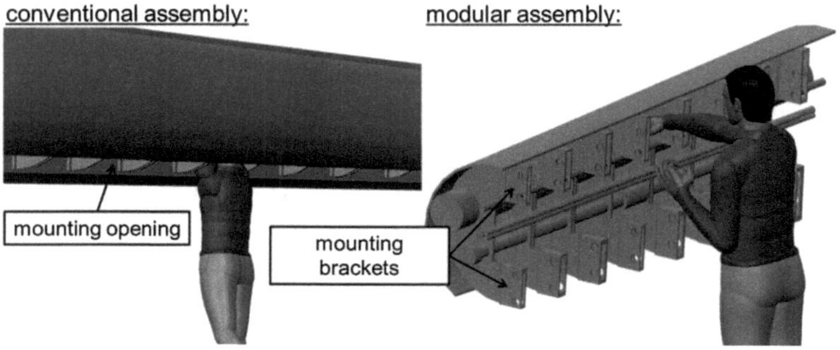

Fig. 2 Comparison of conventional and modular assembly

Instead of standing under the wing to mount parts inside of the leading and trailing edge the assembly can be done under improved agronomical conditions. This assembly concept is comparable to the automobile industry, where e.g. the cockpit is not equipped within the car but preequipped cockpit-modules are used. After the leading and trailing edges are equipped and tested they are assembled to the wing box. The mechanical connection is established by the mounting brackets, in order to connect the electrical harness and the hydraulic pipes secured connectors have to be used.

Technical Feasibility

The verification of the structural strength of a modular wing design is based upon an analytical bending beam model. Through a simplification of the complex geometry of the reference wing to a bending beam with linear decreasing height and depth and with a fixed mounting a direct calculation of the stresses within the wing box was made possible. In contrast to the use of a Finite Element Method (FEM) an analytical bending beam enables a review of different wing geometries, 'cutting' positions, materials, and loadings without creating a complete new model. The following Fig. 3 shows the considered structural loading and geometry of the calculation.

The net loading of an airplane wing is the difference between the lift and the weight force of the wing multiplied with the load factor F. This equals the weight force of the fuselage multiplied with the load factor. As a reasonable simplification, it is assumed that the lift force and the weight force are proportional to the wing depth.

Using the Bernoulli–Euler differential equations for a beam model a calculation of the bending moment and the bending itself is possible [7]. Because of the linear decreasing height and depth of the analyzed wing a numerical Integration via the trapezoidal rule is necessary. For 280 discrete spanwise locations the following values were calculated:

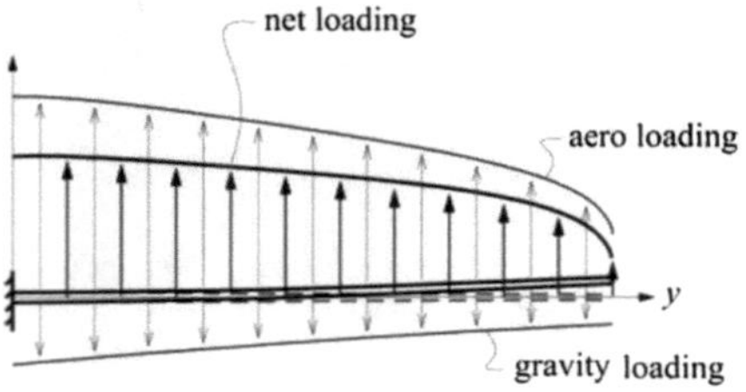

input parameters:

span width	28,00 m
wing depth root	13,60 m
wing depth tip	2,60 m
wing height root	2,30 m
wing height tip	0,24 m
weight fuselage	200.000 kg
load factor F	2,50

Fig. 3 Structural loading and geometry

$$S_{i+1} - S_i = \frac{q_{i+1} + q_i}{2} \cdot (y_{i+1} - y_i)$$

$$M_{i+1} - M_i = \frac{S_{i+1} + S_i}{2} \cdot (y_{i+1} - y_i)$$

$$\theta_{i+1} - \theta_i = \left(\frac{M_{i+1}}{EI_{i+1}} + \frac{M_i}{EI_i}\right) \cdot (y_{i+1} - y_i)$$

$$w_{i+1} - w_i = \frac{\theta_{i+1} + \theta_i}{2} . (y_{i+1} - y_i)$$

with: S = shear force; M = bending moment; θ = bending angle; w = bending; y = y-coordinate of segment; E = elasticity modulus; I = local moment of inertia.

With the boundary conditions, that the shear force S and the bending moment M equals zero at the wing tip and the condition that the bending angle θ and the bending are zero at the root, the bending beam model can be solved. On the basis of the local bending moment and the local geometry of the wing the bending stresses within the top and bottom cover can be calculated.

For the input parameters shown in Fig. 3, a minimal sheet thickness for every discrete element and the use of an aluminum alloy with a yield strength of 350 N/mm^2 a maximal bending moment of 26,560,000 Nm and a maximal bending of approximately 2800 mm was calculated. In addition, the results of this analytical calculation were compared to an FEM simulation. The comparison showed, that the bending beam calculation and the FEM-simulation provide very similar results. As the FEM-simulation takes nonlinear effects into account, the bending calculated with the FEM simulation is always higher than the bending calculated with the bending beam model. For the input parameters shown in Fig. 3 the relative error is 11.5 %. The use of a bending beam model is therefore sufficient for a first evaluation of the technical feasibility of a small-scaled modular wing and is more flexible than a FEM-simulation.

The knowledge of the bending and the material stresses allows a rough dimensioning of the rivet connection between the wing box modules. For example, a connection of two wing box-modules in a distance of 10 m to the wing root would require 2600 rivets with a diameter of 6 mm (calculation according to the DIN 18 800-1 with the parameters shown in Fig. 3 and aluminum rivets). In order to place this 2600 rivets on the cover and taking the minimum rivet distances into account a connection with 5 rivet rows is necessary. The number of necessary rivet rows determines the minimal overlapping of the individual wing box segments and thus allows an estimate of the additional weight as a function of the number of modules. For this example an overlapping of about 96 mm and an additional weight of 7.87 kg was determined.

Besides an analysis of the structural strength, we identified additional factors, which could prevent the realization of a small-scaled modular design: the necessary installation space and the possibility to disconnect electrical cables. Current aviation regulations specify a maximal number of disconnection points of

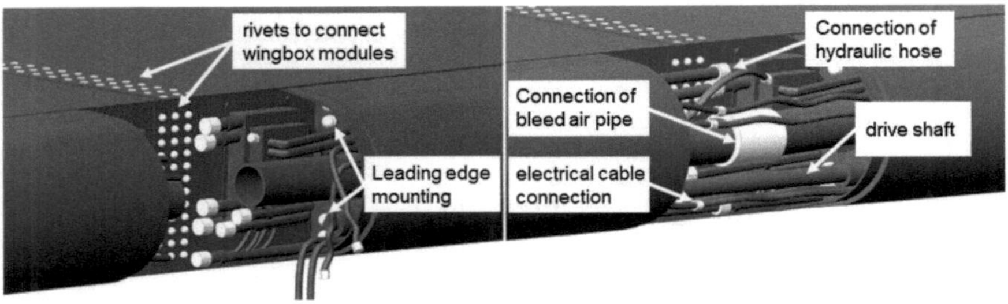

Fig. 4 Installation space leading edge (worst case)

electrical cables. The result is that for a small-scaled modular aircraft wing a complete pre-equipping of all electrical cables is currently not possible.

Furthermore, the analysis of the application example aircraft showed that even at the area of the wingtip there is enough installation space to mechanically connect the trailing edge modules with each other. In contrast to the trailing edge the position with the least installation space for the assembly of the leading edge is not the wing tip. Because the pipes of the bleed air system and the generator cables do not lead from the wing tip to the root the position with the least installation space is just behind the engine mount. Figure 4 shows in an exemplary manner the available installation space for the connection of two modules under the most complicated circumstances.

In this exemplary assembly situation several hydraulic pipes, a bleed air pipe, several parts of the electric harness and a mechanical drive shaft for the slat system have to be connected with each other. The connection is done via an open segment of the leading edge cover which will be closed after the assembly process is finished. The left part of Fig. 4 shows the available space to connect the wing box modules with each other and the space to connect the leading edge to the wing box. Especially the space to assemble the leading edge to the lower right mounting bracket is limited but sufficient. The mounting of internal mounting brackets is done via assembly and maintenance openings at the bottom of the leading edge. The right part of Fig. 4 shows the assembly space to connect the different cables, pipes, harness elements and drive shafts. It can also be seen in this example that the assembly opening must not exceed a maximum size. Although a big opening facilitates the assembly process, it reduces possible positions to secure cables within the leading edge at the same time.

Economic Viability

For a substantial assessment of a small-scaled modular design of an airplane wing the whole aircraft life cycle has to be considered. By an analysis of the whole lifecycle of an aircraft and expert interviews we identified the following 'modularization factors', which are significantly affected by a small-scaled modular design (see Table 1).

Table 1 Modularization factors and possible effects

Modularization factor	Possible effects of small-scaled modular design
Research and development costs	Additional development work; easier subcontracting; use of module catalogs
Manufacturability	Different process capabilities; amount of rework; size of production machines
Difficulty of assembly	Accessibility and weight of components; additional assembly operations
Transport costs	Wider variety of usable transport systems
Fuel and maintenance costs	Fuel burn due to weight and aerodynamics; different maintenance effort

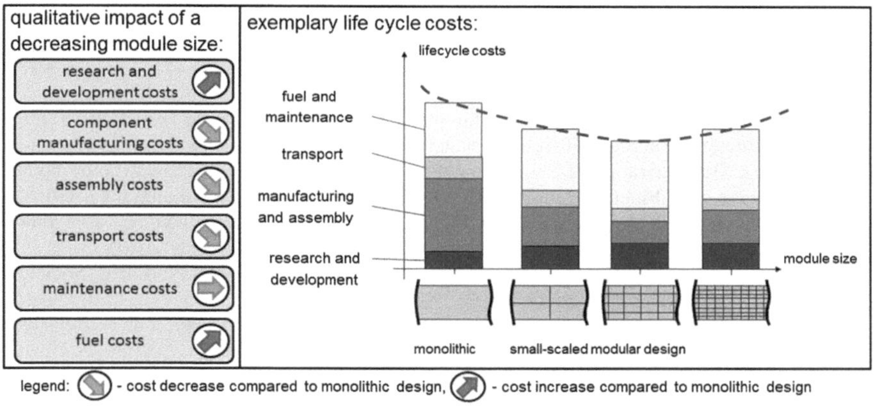

Fig. 5 Lifecycle costs as a function of the module size

Further modularization factors with a lower impact to the lifecycle costs, like ergonomics, the feasibility of retrofitting and a larger product variety, and recycling could be identified, but are not considered within the method to determine the ideal module size. Through this focusing on the most significant modularization factors an application of the developed method in the preliminary design phase is facilitated, as in the preliminary design phase only few and partly imprecise data are available.

The modularization factors include contradictory design targets (see Fig. 5).

In order to facilitate the manufacturing and the transport a small module size and thus a high number of modules should be realized. On the other hand, there are other modularization factors which support a minimization of the number of modules. For example, the fuel burn raises with an increasing number of modules as every interface between two modules causes additional weight and aerodynamic disruptions. Therefore, we use the predicted lifecycle costs as a measurand to determine the ideal module size. By minimizing the total life cycle costs not only one design aspect like the manufacturability is optimized, but a global optimum is reached. In further research activities, the developed cost functions and the set of if then rules should be optimized and described in detail.

Acknowledgments The IPH researched scientific issues and practical challenges in the production of XXL products as part of a joint research project entitled "Innovations for the Manufacture of Large Scale Products." This collaborative project is funded by the Lower Saxony Ministry of Science and Culture and by the Lower Saxony Ministry for Economic Affairs, Labor and Transport. The project was supported by the Investment and Development Bank of Lower Saxony NBank.

References

1. Seo, K.-K. et al.: Approximate estimation of the product life cycle cost using artificial neural networks in conceptual design. Int. J. Adv. Manuf. Technol. **19**(6), 461–471 (Springer Verlag London 2002)
2. N.N.: Intelligent airframe design. Airbus technical magazine, Issue April 2013 http://de.slideshare.net/agromera/a350-fast-speciala350 (June 2013)
3. Bertram, V.: Modularization of Ships, Report within the Framework of Project "Intermodul". http://www.cto.gda.pl/fileadmin/Inne_pdfy/modular_equipment_survey_final.pdf (2013)
4. Blade Dynamics Ltd: A new paradigm in blade technology. Longer blades. More power. Lower cost of energy, http://www.bladedynamics.com/onshore.html (May 2013)
5. Ericsson, A., Erixon, G.: Controlling Design Variants: Modular Product Platforms. Society of Manufacturing engineers, Dearborn (1999)
6. Koeppen, B.: Modularisierung komplexer Produkte anhand technischer und betriebswirtschaftlicher Komponentenkopplungen. Shaker Verlag, Aachen (2008)
7. Munz, C.-D., Westermann, T.: Numerische Behandlung gewöhnlicher und partieller Differenzialgleichungen, 3rd edn. Springer Verlag, Berlin (2013)

Development of Machining Strategies for Aerospace Components, Using Virtual Machining Tools

L. Estman, D. Merdol, K.-G. Brask, V. Kalhori and Y. Altintas

Introduction

The innovation capacity of machining industry is highly dependent on the development of new technologies to improve the efficiency and accuracy in production. It is essential to develop methods for optimal selection of cutting tools, machining strategies and machine concepts for various components and work-piece materials in order to reach the productivity goals. The predictability of machining process is the key to avoid undesired fails like tool-machine collisions, and to reach shorter operation and setup time, energy efficiency, and optimal part quality with respect to dimensional accuracy and surface integrity. Virtual machining tools are built on physics, dynamics, kinematics and control models of machining systems and are of significant importance to support the above activities. These with respect to development of a competitive manufacturing economics to digitally evaluate;

- Process and operation planning, e.g., suitable machining strategies, cutting tools and cutting data, generation of tool paths.
- Preparations for machining and quality evaluation, e.g. preparation of fixturing and clamping devices, preparation of blank, post processing or CNC programming, assembly and presetting of cutting tools etc.

A current study demonstrates the developed cutting tools and strategies in five axis machining of impellers and blades at Sandvik Coromant in partnership with Manufacturing Automation Laboratory at the University of British Columbia. The use of virtual machining tools during process planning is discussed and the industrial benefits are presented.

L. Estman (✉) · K.-G. Brask · V. Kalhori
AB Sandvik Coromant, R&D Application Solutions, 81181 Sandviken, Sweden
e-mail: Linda.Estman@sandvik.com

D. Merdol · Y. Altintas
Department of Mechanical Engineering, Manufacturing Automation Laboratory,
The University of British Columbia, Vancouver, BC V6T 1Z4, Canada

B. Denkena (ed.), *New Production Technologies in Aerospace Industry*,
Lecture Notes in Production Engineering, DOI: 10.1007/978-3-319-01964-2_9,

Machining Strategies

Impeller and blade are among the most complex and costly aerospace parts to machine. The requirements on high productivity and tight component tolerances demand a very careful selection of cutting tools and parameters. High thermo mechanical loads on cutting tools and in cutting zones, excessive distortion due to residual stresses and tool deflection leading to undesired tolerance deviations need to be avoided while reaching a better process efficiency, and lower production costs.

To achieve the mentioned requirements Sandvik Coromant has developed a number of machining strategies including cutting tools, machining methods and machining concepts which are described as follows. The virtual machining tools have been used both for establishing a higher understanding of the cutting process and to verify the chosen strategies reaching extensive productivity improvements. The development of strategies for machining of impeller has been focused generally on applying standard tool concepts such as Coromill Plura for flank milling to achieve a high axial depth of cut to produce the profile in fewer passes. Figures 1 and 2 show the terminology used within process planning for impeller and blade.

Fig. 1 Impeller, Ti6Al4 V, terminology

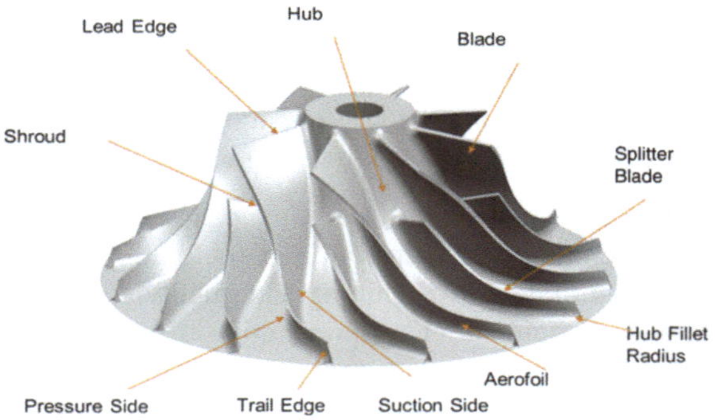

Fig. 2 Blade, X22CrMoV12-1, terminology

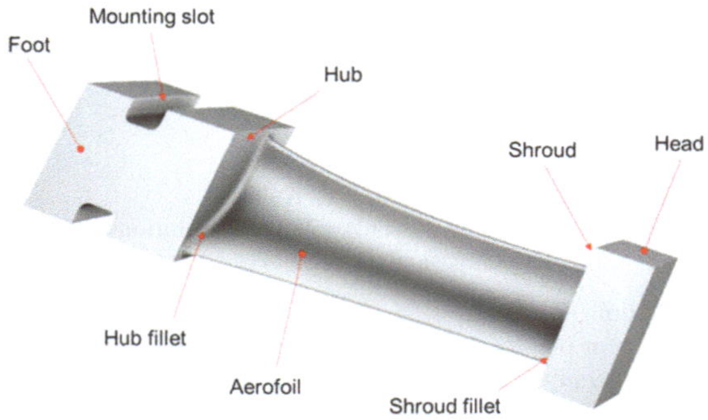

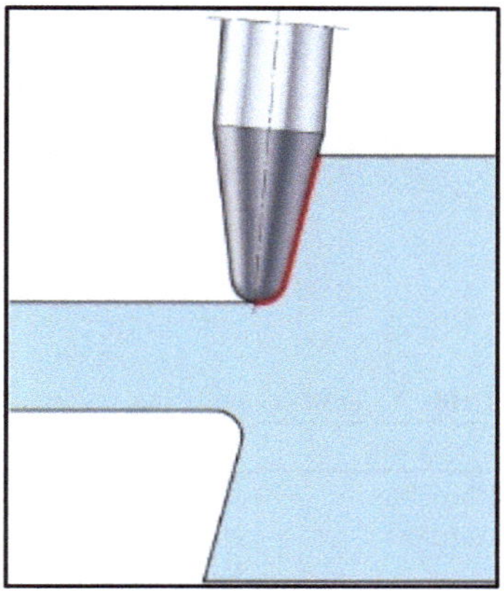

Fig. 3 Flank milling

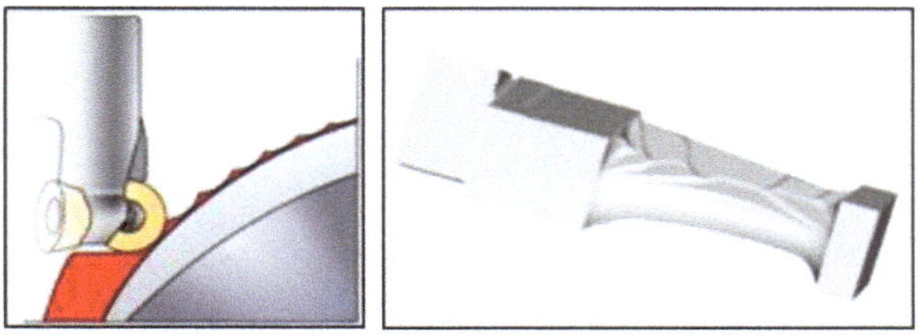

Fig. 4 Waterline milling of Aerofoil

Flank Milling

In flank milling the material is removed with the periphery of tapered ball end mills, see Fig. 3. In order to produce a flank milling tool path, two curves are identified on the design surface. These curves then form the rails on which the tool is rolled.

Waterline Milling of Aerofoil

In this technique, the cutting operation consists of a sequence of 2-dimensional layers with varying depth of cut, each completed before the tool moves down to the next as shown in Fig. 4.

Spiral Milling of Aerofoil (4–5 ax)

Semi roughing operation in which the work (as a helical gear) is given simultaneously a rotary motion and an endways motion, see Fig 5. Sandvik Coromant

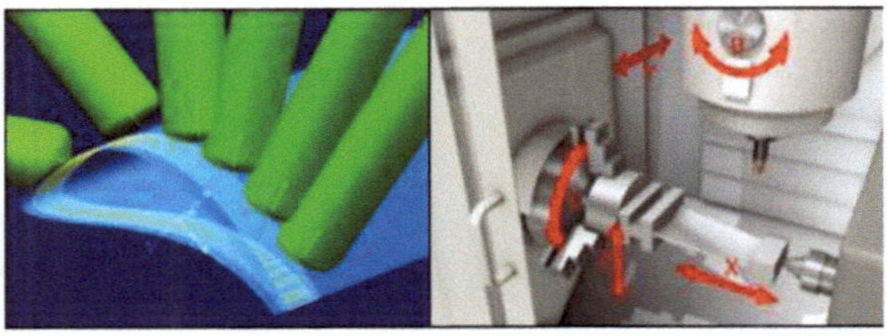

Fig. 5 Spiral milling

Table 1 Timesaving using the machining strategies developed at Sandvik Coromant

Operation	Machining	Conventional	min:sec	Recommended	min:sec
Rhombus	Roughing	Face milling	07:00	Face milling	04:25
Aerofoil	Roughing	Face milling	08:50	Waterline milling	07:00
	Semi-finishing	Spiral milling	02:40	Spiral milling	02:12
	Finishing	Spiral milling	06:00	Spiral milling	05:00
Hub and shroud	Roughing	Point milling	09:20	Flank and point milling	05:50
	Finishing	Point milling	20:00	Flank and point milling	09:15
Head and footh	Finishing	*	02:00	Face milling	02:00
Mounting slot	Roughing and finishing	*	01:40	Slot milling	01:40

application knowledge in combination with product development of the holding system of CM600 has been leading the development of this methods.

Flank Milling of Hub and Shroud

The conventional method of machining hub and shroud is point milling where the cutting is done with the bottom edge of the ball ended cutter. A 3D-profile is generated with successive and many levels of passes.

The combination of point milling and flank milling has reduced the cycle time of this operation from 20.00 to 9.15 min, see Table 1.

Optimization

Using standard commercial CAM packages to create an optimized tool path usually results to an undesired variation of hex. In this section the use of optimization platform MACHpro developed at The MAL University of British

Columbia is discussed. The platform is used to verify the developed strategies and to reach among other a constant chip load (hex) optimized cutting loads, etc. The software analyzes the NC file and then, based on the stock removal simulation process which is integrating physics-based material models, calculates the real amount of material to be removed for each position of the cutter along the tool path. Along the tool path the software can optimize the machining/cutting data with respect to:

- Cutting force
- Chip thickness
- Spindle torque/power
- Tool deflection

For example, it is possible to determine the required cutting forces along the tool path, with respect to the amount of material to be removed. In areas where there is a reduced amount of material to be removed and cutting conditions are favorable, the software will automatically increase the feed rate and update the NC file. The software does not rely on amount of material removed to calculate process outputs like forces. It rather uses intersection surface between tool and work piece to calculate actual chip distribution along helical flutes and combines this information with complex material models to determine the instantaneous load on the tool. MACHpro can also detect the presence of chatter; feed drive loads and allows the detailed analysis of process states at crucial states of the tool path while considering the CNC dynamics of the machine which has great influence on chip loads and cycle times, see Table 2.

Table 2 Optimized cycle time for Impeller machining using MACHpro

Operation	Cycle time		Productivity improvement (%)
	Application strategy (min:sec)	NC-code Optimization (min:sec)	
Roughing 1	02:40	02:18	14
Roughing 2	01:47	01:23	22
Roughing 3	01:48	01:35	12
Roughing 4	01:10	01:00	14
Roughing 5	01:11	01:03	11
Semi finish 1 (Blade)	00:54	00:37	31
Semi finish 2 (Blade)	01:30	01:06	27
Semi finish (Hub)	02:59	01:34	47
Time Saving/blade	13:59	10:36	24
Total Time Saving (eight blade Impeller)	111:52	84:48	24

Conclusions

Machining strategies, cutting tools, and application methods developed at Sandvik Coromant have led to significant productivity improvement and component quality assurance. The use of virtual machining software MACHpro as a natural part of production planning process has resulted to additional cycle time reduction and tolerance accuracy when machining costly and complex to machine parts like impeller and aerospace blades.

Influence of 5-axes-kinematics Geometrical Accuracy in Riblet Manufacturing Processes

Berend Denkena, Jens Köhler and Thomas Krawczyk

Abstract The effectiveness of gas turbines can be improved significantly by decreasing the friction losses. Compared to smooth surfaces riblet-structures have been proven to reduce skin friction in turbulent flow up to 10 %. For the technical application on compressor blades in turbo machines, micro riblet structures with a riblet width between 20 and 120 μm and a depth of the half width are required. Furthermore, the structuring process must be able to generate micro patterns, which are orientated in stream direction, on free formed surfaces. So far, riblet structures have been ground on single curved NACA6510. In order to grind riblet structures a 5–axes-grinding process is required. However, the complex motion of the grinding wheel results in defects on the ground geometries. This paper presents the effect of the 5-axes-kinematics on the riblet aspect ratio and, as a consequence, general conditions, which enable a grinding process without defects on the ground riblets.

Keywords Five axes grinding · Micro pattern · Dressing

Introduction

A major aim in turbomachinery design is the increase of efficiency to reduce the fuel consumption of aircraft engines or gas turbines and hence the operating costs and CO_2-emissions. To attain an increase of efficiency flow losses have to be reduced. A large proportion of flow losses is generated by skin friction between the blade surfaces of the axial compressor and the working fluid [1]. Hence, a decrease of skin friction leads to an additional increase of efficiency. It is known that small ribs on the

B. Denkena · J. Köhler · T. Krawczyk (✉)
Institute of Production Engineering and Machine Tools, Leibniz Universität Hannover,
An der Universität 2, 30823 Garbsen, Germany
e-mail: krawczyk@ifw.uni-hannover.de
URL: www.ifw.uni-hannover.de

B. Denkena (ed.), *New Production Technologies in Aerospace Industry*,
Lecture Notes in Production Engineering, DOI: 10.1007/978-3-319-01964-2_10,

surface oriented in the mean flow direction can reduce skin friction in the case of turbulent flow when their geometric features are appropriate to the local flow conditions. Ideal riblet structures on compressor blades should have a riblet width between 20 and 120 μm and an aspect ratio of riblet height to width of 0.5 [2, 3].

Compared to manufacturing processes such as laser machining [4], EDM, micro milling and micro planing [5, 6] grinding with multi-profiled wheels has been established as an effective method for generating riblet structures on large scale surfaces. So far, riblet structures with an aspect ratio of 0.5 and a width of 60 μm have been ground on single curved NACA6510 profiles whereby a reduction of the near wall friction of about 4 % has been achieved [7, 8]. However, real compressor blades have double curved or free formed surfaces. These surfaces have to be machined with a 5-axes-grinding process. Hereby additional requirements on the grinding process have to be considered [9, 10]. The tool paths have to be curved in order to follow the curved stream flow. Such complex tool paths were used for e.g. belt-grinding in mould manufacturing [11]. However, this grinding belts were not profiled, hence the influence of the 5-axes-kinematic on the ground riblet structures is unknown.

To close the research gap in structuring of free formed surfaces the relevant influencing factors on the accuracy of the riblet-geometry in 5-axes-grinding of riblet structures on double curved compressor blades are investigated. This paper deals with the impact of the 5-axes-kinematics on the ground riblet geometries, the resulting boundaries and strategies, which enables a shift these boundaries.

Materials and Methods

In order to investigate the influence of the 5-axes-kinematics on the riblet geometries grinding simulations were done which the simulation system CutS®. CutS® used a three-dimensional dexel model, which represents the workpiece and calculated the material removal due to the interaction with the tool. The verifications of the simulations have been carried out on a Blohm Profimat 407 5-axes-surface grinding machine. The dressing system, using a diamond profile dressing roller, is integrated into the machine. Vitrified bonded wheels whit a grain sizes of 17 μm have been selected due to their good dressability and good profile holding. The grinding wheel profiles were dressed by the shift strategy. Here, one flank of the profiles is dressed by a first plunge movement. The second plunge movement is carried out with an axial offset of the dressing roller, whereby the other flank of the profiles is dressed. Due to the process kinematics it is possible to generate all of the profiles within two plunge movements. In order to avoid the influence of tool wear during the verification Obomodulan® has been choosen as workpiece material. Obomodulan® based polyurethane and has a density of 1200 kg/m^3, compressive strength of 82 MPa and a bending strength of 94 MPa.

Results and Discussion

By grinding curved riblets undercuts can emerge, which result in a damage of the riblet structures. The undercuts are a result of the contact conditions of grinding wheel and workpiece. When a grinding wheel moves on a curved tool path, the target groove geometry should be equal in cross section A–A and B–B at the time t = 0 (Fig. 1). Due to the straight contact length, a groove at the right side of the target groove will be ground. When the grinding wheel moves forward on the curved tool path to the time t = n an undercut results (Fig. 1, top left). As a consequence the width of the ground groove will be enlarged and the face angle of the groove differs to the targeted angle. To investigate the influencing factors on the undercut and to define the geometry of curved riblet structures, which can be ground on free formed surfaces, the divergence of target face angle and generated face angle was calculated as a function of tool path radius for different grinding wheel diameter.

Figure 2 illustrates the divergence of the target face angle alpha and generated face angle for varying grinding wheel diameter R_{WZ}. The results show that undercuts can be avoided when the diameter of the grinding wheel diameter is adjusted to the tool path radius R_{path}. By applying a grinding wheel with a diameter of $R_{WZ} = 40$ mm the divergence angle decreases when the tool path radius rise till a radius of 60 mm. For bigger tool path radii the divergence angle does not change. The minimal tool path radius, which does not result in undercuts, is independent of the grinding wheel diameter. For a grinding wheel diameter of $R_{WZ} = 144$ mm no difference between the target angle and the actual groove angle from a tool path radius about 210 mm occurred.

Similar results were observed in grinding experiments. The green line in Fig. 2 shows the divergence angle of target grooves and the ground one. Here, a grinding wheel with a V-profile, a profile angle of 60° and a diameter of 144 mm was applied

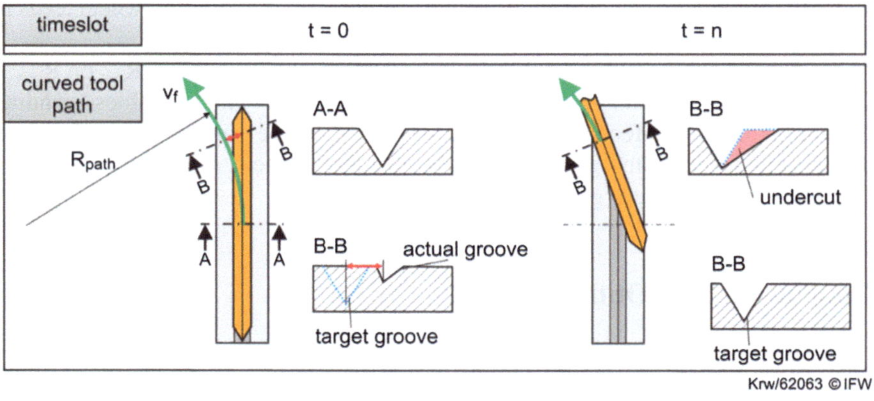

Fig. 1 Emergence of undercuts

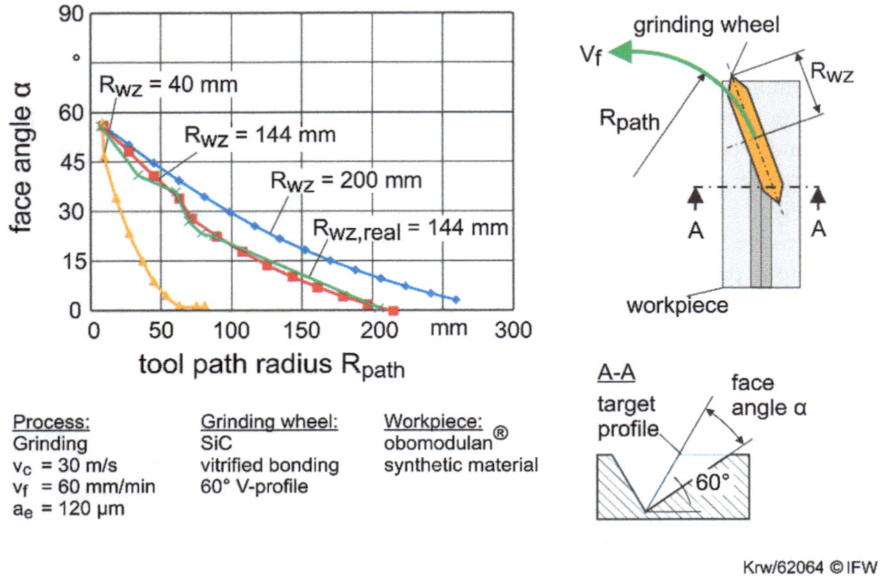

Fig. 2 Relationship of undercuts and tool diameter by circular motion R_{path}

and the same divergence angle like the simulated one results (green and red line in Fig. 2). To conclude the use of a grinding wheel diameter which is adapted on the required tool path radius enables grinding of curved riblet structures.

In order to avoid the influence of tool wear on the riblet geometries grinding wheels with a big diameter should be used. However, this grinding wheel cannot realize small tool path radii without geometrical defects. Thus, a new grinding strategy has been investigated to realize small tool path radii and big grinding wheel diameter. Until now, the grinding wheel are normal to the workpiece surface, which results in a straight contact length. By tilting of the grinding wheel about the angle β an adjusted position of the profiles on the grinding wheel leeds to more curved contact length (Fig. 3, left).

On the right part of Fig. 3 the influence of grinding wheel radius R_{wz} on R_{crit} is shown. R_{crit} is the smallest possible tool path radius which enables grinding without geometric defects for different tilt angles. These simulation results show that grinding with tilted grinding wheels enables smaller critical tool path radii and/or the use of bigger grinding wheels.

Conclusion and Outlook

Ideal riblet structures reduce the wall near friction up to 10 %. Industrial application requires grinding of micro patterns with an aspect ratio of 0.5 on free formed surfaces in a shape accuracy about 10 µm. The simulation and grinding

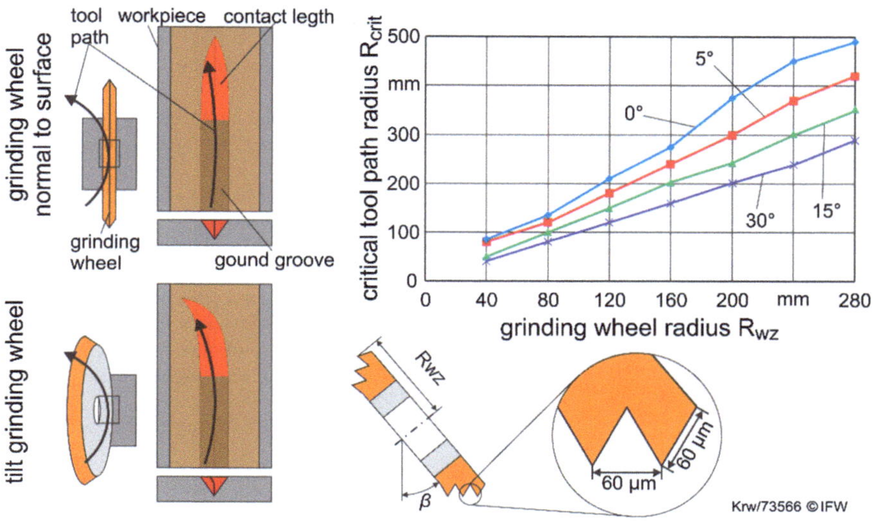

Fig. 3 Grinding with tilt grinding wheels

experiments show, that due to the 5-axes-kinematics defects on the ground riblet structures can occur. However, if a grinding wheel moves on a circular path, the ground structure are not influenced if a tool path radius above a critical tool path radius is used. This critical radius can be decreased by applying smaller grinding wheels. Moreover, a tilted grinding wheel with an adapted profile geometry reduces the critical tool path radius, too. As future work, these results will be verified in grinding investigations. More over, current investigations aim to quantify the friction reduction by riblets in a real compressor.

Acknowledgments The investigations described in this paper were undertaken with the support of the German Research Foundation (DFG) within the project "Riblets on Compressor Blades" (DE 447/51-3).

References

1. Gümmer, V.: Pfeilung und V-Stellung zur Beeinflussung der dreidimensionalen Strömung in Leiträdern transsonischer Axialverdichter, Fortschritt-Berichte VDI Reihe 7 Nr. 384, VDI Verlag, Düsseldorf (2005)
2. Ninnemann, T., Ng, W.-F.: Loss Reduction Using Riblets on a Supersonic Through-flow Fan Blade Cascade. Trans. ASME, Ser. I J. Fluids Eng. **126**(4), 642–649 (2004)
3. Bechert, D.-W., Bartenwerfer, M.: The viscous flow on surfaces with longitudinal ribs. J. Fluid Mech. **206**, 105–129 (1989)
4. Siegel, F., Klug, U., Kling, R., Ostendorf, A.: Extensive micro-structuring of metals using picosecond pulses - ablation behavior and industrial relevance, In: Proceedings of LPM 2008,

9th International Symposium on Laser Precision Manufacturing, 17-20 Juni, Quebec, Canada, (2008)

5. Fischer, S.: Fertigungssysteme zur spanenden Herstellung von Mikrostrukturen, PhD thesis, RWTH, Aachen (2000)
6. Weinert, K., Blum, H., Jansen, T., Rademacher, A.: Simulation based optimization of NC-shape grinding process with toroid grinding wheels. Prod. Eng. Res. Dev. (WGP), **1** (3), 245,252, (2007)
7. Uhlmann, E., Piltz, S., Doll, U.: Funkenerosion in der Mikrotechnik. Einsatzgebiete und Verfahrensgrenzen. Werkstatttechnik wt-online, **12**, 733–737 (2004)
8. Denkena, B., Köhler, J., Wang, B.: Manufacturing of functional riblet structures by profile grinding. CIRP J. Manuf. Sci. Technol. **3**(1), 14–26 (2010)
9. Lietmeyer, C., Denkena, B., Kling, R., Krawczyk, T., Overmeyer, L., Reithmeier, E., Scheuer, R., Vynnyk, T., Wojakowski, B., Seume, J.R.: Recent advances in manufacturing of riblets on compressor blades and their aerodynamic impact. J Turbomach. **135** (4), 2013
10. Brinksmeier, E., Riemer, O., Osmer, J.: Tool path generation for ultra-precision machining of free-form surfaces. Prod. Eng. Res. Dev. **2**, 241–246 (2008)
11. Tönshoff, H.K., Denkena, B., Böß, V., Urban, B.: Automated finishing of dies and molds. Prod. Eng. Res. Dev. **9**(2), 1–4 (2002)

New Technology for High Speed Cutting of Titanium Alloys

Eberhard Abele and Roland Hölscher

In the aviation industry the weight of an aircraft has a strong influence on the operating costs. Less weight means on the one hand less energy consumption and on the other hand a bigger capacity for passengers and freight. Assuming a Boeing 747 would carry 10 % less fuel, a payload of 10 tons would be feasible. Hence the net load ratio with equal weight is drastically improved. Furthermore secondary savings such as lowering the demand of fuel and reducing maintenance costs would promote a general reduction of the total costs [1].

To achieve such savings, the usage of lighter materials such as carbon fiber reinforced plastics (CFRP) and titanium alloys is increasing. More CFRP means not only weight saving, but also an increase of the usage of titanium. A reason for that is a high electrochemical potential difference between CFRP and aluminum alloys, which enhances corrosion. The connection between CFRP and titanium alloys reduces such potential difference by 80 %. Thus the usage of CFRP leads to a suppression of aluminum parts in the aviation industry. Fewer metal and fewer sandwich materials means less metal machining. The new material mix of the Boeing 787 Dreamliner and the Airbus A350 WXB is shown in Fig. 1. The fraction of titanium alloys in the new Boeing and Airbus tripled compared to previous aircrafts.

Even though the metal removing volume generally decreased in the aerospace industry—the machining time increased. A reason for that is the very low and viable titanium removal rate. Over the last decade such a removal rate quintupled with aluminum parts; whereas the removal rate with titanium parts doubled (Fig. 2, left). Today a metal removal rate of 15 l/min can be attained by machining of aluminum. In the field of titanium machining the rate is about 0.25 l/min with

E. Abele (✉) · R. Hölscher
Institute of Production Management, Technology and Machine Tools (PTW), Technische Universität Darmstadt, Petersenstr. 30, 64287 Darmstadt, Germany
e-mail: abele@ptw.tu-darmstadt.de

R. Hölscher
e-mail: hoelscher@ptw.tu-darmstadt.de

B. Denkena (ed.), *New Production Technologies in Aerospace Industry*,
Lecture Notes in Production Engineering, DOI: 10.1007/978-3-319-01964-2_11,

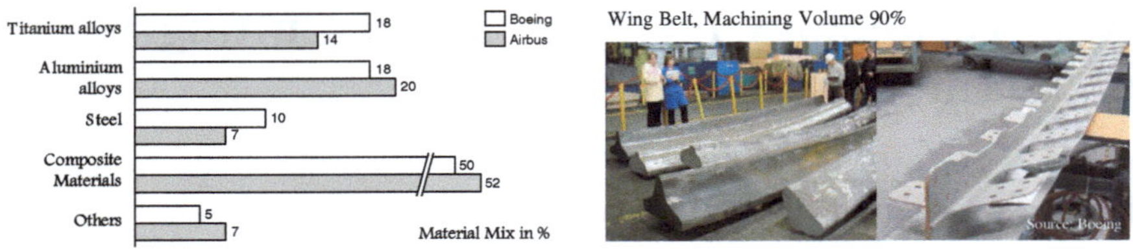

Fig. 1 New materials mix in next aviation generation (*left*), aviation structure part (*right*)

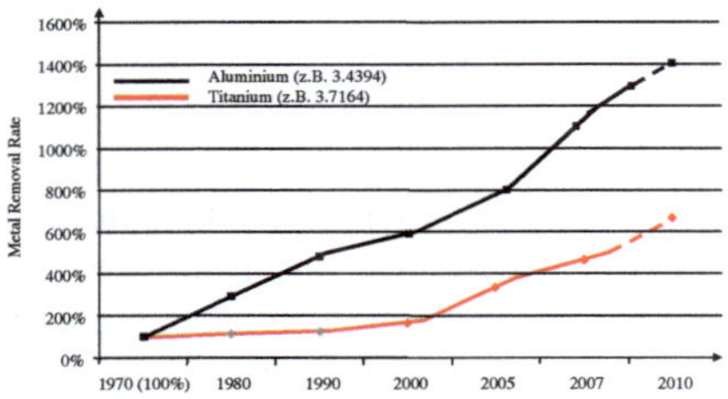

Properties	Ti6Al4V	Ck45	AL 7075
Specific mass (g/cm^3)	4,43	7,78	2,8
Young's Modulus (GPa)	110	210	72
Tensile strength R_m (MPa)	900-1180	630	550
Elastic Limit $R_{p0,2}$ (MPa)	830-1030	390	470
Elongation A5 (%)	8-15	14-17	12
Hardness (HB)	320	170	55-163
Heat Capacity c_p (J/kgK)	520	486	862
Heat Conductivity λ (W/mK)	7,12	51,9	115-140

Fig. 2 Growing material removal rate of aluminum and titanium alloys (*left*). Properties of Ti6Al4V, Ck 45 and Al7075 (*right*)

solid carbide tools. In consequence the machining costs for titanium parts are extremely high compared to aluminum parts [2].

In the following the paper presents the reasons why the metal removal rate in titanium is still very low. The low E modulus, the high cutting forces and the very low heat coefficient are tough on the machining process (Fig. 2, right). A very low heat conductivity of $\lambda = 7.12$ W/mK of Ti6Al4V (in comparison with Ck 45 and Al 7075), causes along with cutting speed, feed and width of cut a development of extreme temperatures on the cutting edge.

The heat conductivity of aluminum is greater than that of titanium by a factor of 15. Another major difference lies in the heat dissipation in the chips. Over 75 % of the process heat is dissipated in the chips during aluminum machining whereas only 25 % goes into the chip in titanium machining. That means the tool faces an enormous temperature load. High temperatures at the cutting edge cause adhesion and diffusion processes, leading to high tool wear (Fig. 3).

A low E modulus also shows that titanium yields to the pressure of the cutting material. The danger of vibration behavior with weak parts increases. This is further supported through yield strength ratio and only partially allows plastic deformation. Because of the compliant material, the material springs back further under the impact of the cutting force, causing a decrease of the clearance angle in the area of the clearance surface. Low cutting velocities and high forces combined with a low excitation frequency may cause chatter [3].

The extremely low heat coefficient causes a disability of the High Speed Cutting (HSC)-Effect with titanium machining. The increased tensile strength of titanium at higher cutting speed and the missing ability of the introduction of the

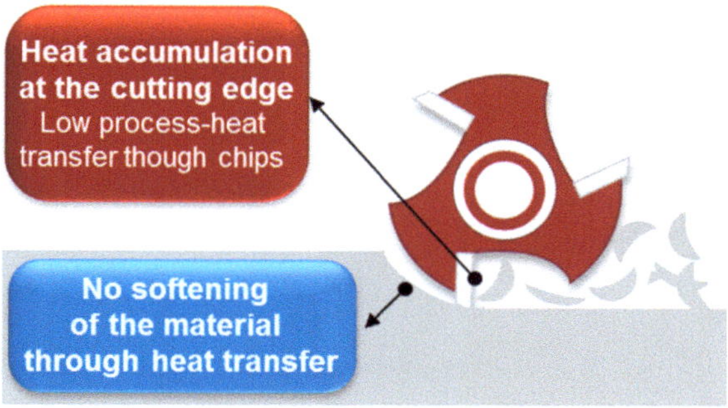

Fig. 3 Heat accumulation at the cutting edge

cutting heat cannot cause a decrease in the cutting forces and also result not in material softening. The titanium material cannot transfer the process heat through the chips, causing heat accumulation and thermal overload at the cutting edge [4].

To use the advantages of the HSC-Effect in titanium machining the conditions have to be artificially established. The warm machining process offers promising starting points. In the conventional titanium machining the goal was a high cooling efficiency, whereas now additional energy in form of heat is transferred into the process. The aim is to achieve a better machining, by influencing the material characteristics of the machining process [5].

The material properties of Ti6Al4V at growing temperatures shows that tensile strength, shear strength, E modulus, tendency of hardening, heat conductivity and specific heat capacity decreases. This change of the material properties means for the machining process that the cutting forces decrease and also the wear of the cutting tools decrease. The tool life increases and the chip formation improve (Fig. 4) [6].

The warm machining offers completely new potentials as to the changed material characteristics. Because a multiplicity of titanium parts are structure parts with big dimensions, the heat transfer is of crucial importance. Figure 5 gives an overview on the available possibilities, structuring them according to "through and partial heating" of the part [7].

Theoretically only the area that is going to be machined on can be heated which can be realized through a laser beam that is nowadays relatively cheap and applicable at different power settings. Most obvious advantages of such a heating

Material focused effects	**Process focused effects**
• tensile strength → decreases	• cutting forces → decreases
• shear strength → decreases	• wear of cutting tool → decreases
• E modulus → decreases	• tool lifetime → increases
• tendency of hardening → decreases	• chip formation → improves
• heat conductivity → increases	• milling temperature → decreases
• spezific heatcapacity → increases	• cutting rate → increases

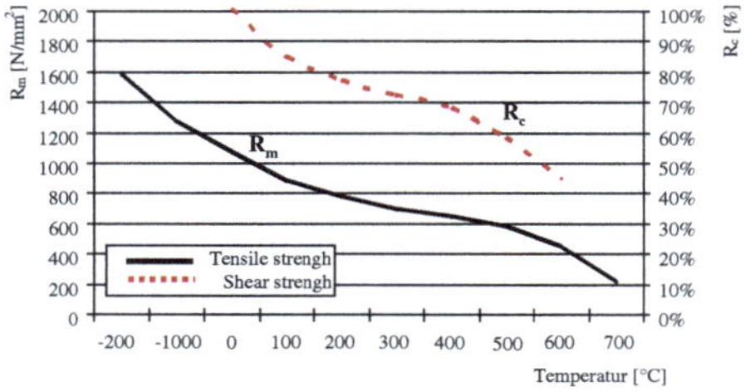

Fig. 4 Material and process effects of warm machining (*left*). Tensile and shear strength decreases at high temperatures (*right*)

Fig. 5 Options for heating the titanium alloy

Heating up the hole part	**Heating up only the cutting area**
• usage of residual heat of force	• heating through friction
• heating in oil	• heating through resistance (AC/DC-current, radiofrequencies)
• heating using electric-furnace	• arcload
	• microwaves
	• plasma beam
	• laser beam
	• UHM Technology
	• elektomagnetic induktion

source is that it is able to be positioned and moved easily, which comes in handy with the processing of more complex parts. During the heating process with a laser beam only the part's surface (max. depth 1 mm) is heated. The disadvantage of the usage of such a device is the partial heating of the surface due to heat power loss. The heat can only be transferred through heat conductance. A low heat coefficient of Ti6Al4V is not giving the desired support. Hence the heat dissipation is not homogenous and the temperature drops rapidly from the surface toward the center of the part. That is a big disadvantage when trying to realize improvements in metal removal rate. A further analysis of the manufactured part's surface is crucial to figure out if the heat input caused any unwanted structural damage or induced any internal stress [8].

Another possibility to achieve the warm machining process is the energy feeding input through induction, which can introduce the heat into the material up to 15 mm due to a relatively homogenous temperature profile. In front of the tool/spindle the inductor is running which is heating the work piece. The process is relatively easy adjustable in terms of geometry and positioning of the spindle. The depth of the inductive generated fields paired with the depth of the heat insertion can be adjusted by the frequency. Through the feed rate of the machine process and the desired temperature difference, along with the specific heat and the heated mass, the demand on the heating source can be calculated. This however means that with the power request of the generator, different temperatures in the part which will machine can be step less adjusted.

Figure 6 shows on the left side the experimental setup. The inductor which is arranged before the tool is heating the material. On the right side the inductor is shown in detail. The marked area is faced to the work peace surface during milling. This part is directly related to the preheated work surface. In this case, the inductor geometry has to be adapted to the machining operation. With the help of the booster material the magnetic field is limited on a certain area on the work piece surface.

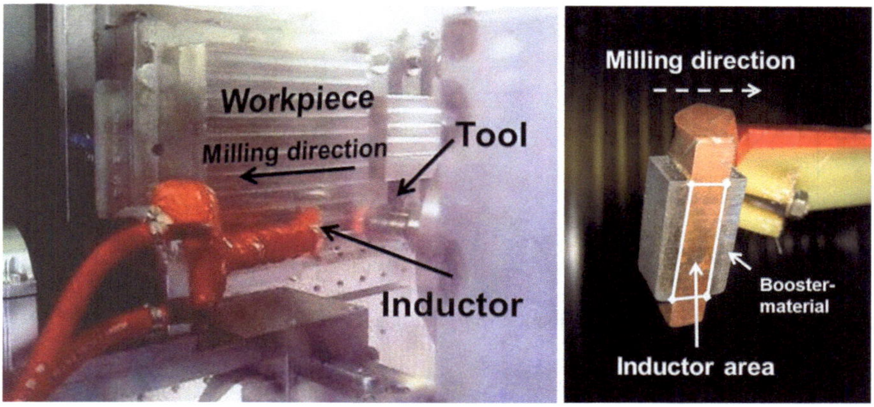

Fig. 6 Test setup (*left*) and Inductor (*right*)

The main focus of research deals with finding of ideal cutting temperatures as well as technological parameters to increase the productivity. Initial investigations showed a decrease in cutting forces by about 30 %. The results show, that tool life was the same compared to conventional machining.

Talking about preheating machining it should be mentioned, that the material removal rate must be adapted to conventional machining. Due to the state of the art the material removal rate shall beat the value for peripheral milling cutter (cutting speed $v_c = 70$ m/min, feed per tooth $f_z = 0.3$ mm, $a_e = 0.5 * d$ mm, $a_p = 10$ mm, $z = 5$). In addition to conventional machining the tool life expected for preheating titanium alloys shall beat 45 m with uncoated carbide cutting inserts and 110 bar inner cooling and additional flood cooling with ester oil. Regarding the conventional machining or HPC-Strategy, the new HSC-Strategy requires increased cutting speeds. The aim of the new technology implementation is to achieve higher productivity. Thus the material removal rate can be raised.

Figure 7 shows the measured temperatures at different depths in the work piece. For this purpose thermocouples were inserted laterally into the workpiece material. They measured the temperature at various depths. (The sensors were inserted 3, 6, 9 and 12 mm in the titanium alloy work piece at a distance of 20 mm.) The induction system produced 60 KW and a frequency of 18 kHz, which the inductor transfers in heat at the follow step. Figure 7 (left) shows the realized temperatures from about 680 °C at 3 mm depth up to 280 °C in 12 mm depth, which were produced by the induction. The measured data converge over the time, when the temperature is allocated in the test material by conduction. During machining the milling tool follows the inductor. Thus the preheated material is machined immediately. Through the effective time which the inductor is heating, this can be controlled by the feed rate or by the power of the induction system, so that the temperature can be controlled. Figure 7 (right) shows a higher material removal with the adjusted technology parameters ($v_c = 300$ m/min, $f_z = 0.1$ mm $a_e = 0.5 * d$ mm,

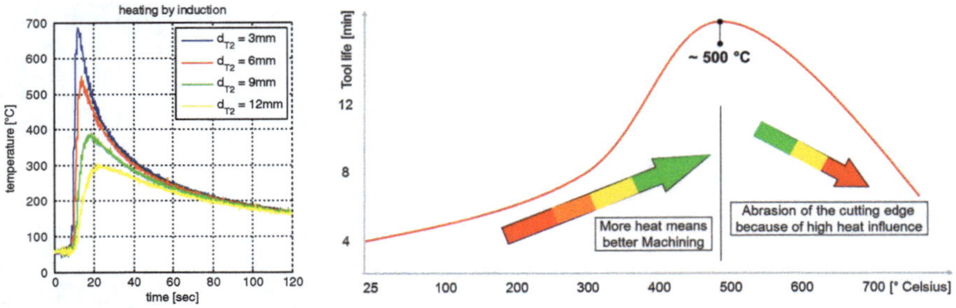

Fig. 7 Induced temperature distribution (*left*) and ideal cutting temperatures (*right*)

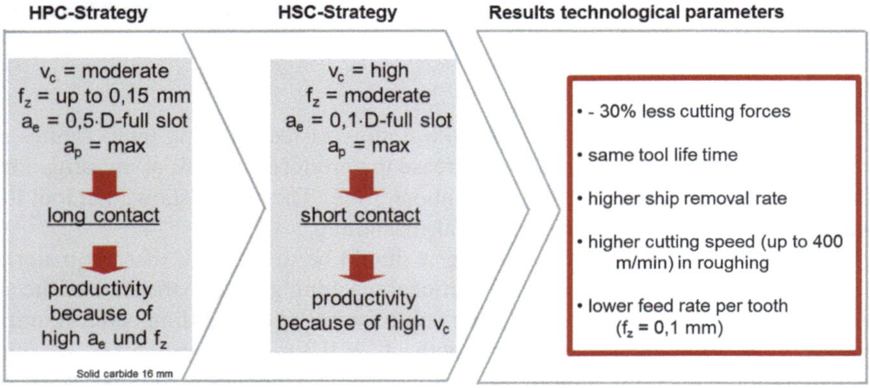

Fig. 8 Warm machining of titanium alloys require a different machining strategy

a_p = 10 mm, z = 5) in comparison with the conventional machining. The best machining results are reached at temperatures around 500 °C.

The investigations showed that the cutting speed could be significantly increased with the preheating machining. In addition, the results showed that there is a possible way to implement an HSC-Strategy in machining titanium alloys (Fig. 8). In future, further investigations must show whether the surface of the titanium alloys has structural changes resulting from preheating or induces unwanted stresses. If the work piece surface does not confirm the requirements or the preheating modified the tolerances cause of thermal draft, there could be a solution through a conventional finishing process.

Conclusively warm machining of titanium alloys at temperatures around 500 °C allows lower cutting forces and less tool wear. In addition the process of the induction assisted machining allows to reduce the process time through the increase of technological parameters.

Hence production costs of titanium parts, which have a high removal rate, are going be noticeably lowered.

References

1. Abele, E., Hölscher, R.: Ein Leichtmetall macht's den Zerspanern schwer, Werkstatt und Betrieb, WB 7-8/2009, pp. 46–51. Carl Hanser Verlag, München (2009)
2. Peters, M., Leyend, C.: Titan und Titanlegierungen, pp. 259–274, 351–367. Wiley, Weinheim (2002)
3. Narutaki, N., Murakoshi, A., Motonishi, S., Takeyama, H.: Study on machining of titanium alloys. Ann. CIRP **32**(1), 591–634 (1983)
4. Rahman, M., Wong, Y.S., Zareena, A.R.: Machinability of titanium alloys. JSME Int. J. Ser. C **46**(1), 107–115 (2003)
5. Eugene, R.L.: Tooling structure: interface between cutting edge and machine tool. Keynote papers WSUD USA. Ann. CIRP **49**(2), 591–634 (2000)
6. Ginta, T.L., Nurul Amin, A.K.M., Lajis, M.A., Mustafizul Karim, A.N., Mohd Radzi, H.C.D.: Improved tool life in end milling Ti-6Al-4V through workpiece preheating. Eur. J. Sci. Res. **27**(3), 384–391 (2009)
7. Ginta, L.T., Lajis, M.A., Nurul Amin, A.K.M.: The performance of uncoated tungsten carbide insert in end milling titanium alloy Ti-6Al-4V through work piece preheating. Am. J. Eng. Appl. Sci. **2**(1), 147–153 (2009)
8. WL 3.7164-100. Werkstoffleistungsblatt Band 2. Luft- und Raumfahrt; Werkstoff-Handbuch der deutschen Luftfahrt. Deutsches Institut für Normung (DIN), Berlin, Germany (1992)

Cutting Lightweight Materials with Surface Modified Tools

Frank Barthelmä and Heiko Frank

Abstract The processing of new and advanced high performance materials becomes more important due to new cutting materials, adjusted tool geometries and coating systems. For the development of innovative tools and technologies, new approaches and procedures were used with consideration of the requirements and parameters of the cutting material [1]. The optimization of cutting edges and hard coating systems is one possibility to improve the process productivity and the product quality of produced parts. Practical examples from research and development, and the practical transfer of research results into the industrial practice will be presented in this publication.

High Performance Materials and Fields of Applications

The demand for weight reduced parts and products in the different application fields as aircraft or automotive industry as well as the mechanical engineering leads to the substitution of the normally used steel or cast materials by aluminum or high performance materials. In the aerospace industry titanium and titanium alloys become more important. These functional lightweight materials often are difficult to machine. Based on these challenges, extensive investigations must be carried out. Selected examples of difficult to machine materials will be given in the next sections. Of special interest are titanium and titanium alloys and carbon respectively glass fiber reinforced plastics (CFRP/GRP) for the application in the aircraft industry.

Examples are the threading as well as the milling of titanium based alloys. A further key aspect is the drilling and milling of CFRP/GRP with regard of

F. Barthelmä (✉) · H. Frank (✉)
Society for Production Engineering and Development (GFE), Schmalkalden, Germany
e-mail: f.barthelmae@gfe-net.de

H. Frank
e-mail: h.frank@gfe-net.de

B. Denkena (ed.), *New Production Technologies in Aerospace Industry*,
Lecture Notes in Production Engineering, DOI: 10.1007/978-3-319-01964-2_12,

functional adjusted cutting edge microstructure, development of new coating systems and adapted surface post-treatment processes.

Selected Cutting Examples: Titanium and CFRP/GRP Materials

Cutting of Titanium

At first, systematic investigations were done to develop cutting strategies to thread in titanium based alloys. Here the objective was the determination of the optimized parameters for the tool macro- and micro geometry, the proper wear resistant coating and the post-treatment process after coating deposition.

First the characteristically wear behavior of hard coating systems against titanium material was determined in tribological tests at room temperature and elevated temperatures. Conventional coatings like TiAlN or TiAlCN shows a strong adhesion of titanium material, in particular at elevated temperatures. Using diamond like carbon (DLC) coatings with a very low friction coefficient in the tribological tests, the adhesion and the abrasive wear can be reduced.

Cutting edge preparation and pre-treatment also strongly influences the cutting process and the friction behavior. The pre- and post-treatment were done with drag finishing process, whereby a cutting edge radius of about 5 μm shows the best results in the cutting tests.

The combination of pre- and post-treatment and the DLC coating leads to a significantly increase of more than 60 % in the life time of DLC coated tapping tools compared to the standard tools. This increase was verified with different process parameters (Fig. 1). Furthermore the low friction coefficient of the used DLC coating can be used to reduce process temperatures and cutting forces and moments [2].

A similar procedure was used to improve the tool life in high performance milling of the titanium alloy TiAl6V4. Cutting edge preparation and post-treatment were used for deburring and stabilizing the cutting edge, the optimized cutting edge was about 5 μm. Post treatment has an effect on the friction behavior and the surface roughness [3].

Milling of TiAl6V4 leads to a higher process temperature, therefore DLC coating are not suitable due to an insufficient thermal stability. In the investigations, a combination of oxides and nitrides in one hard oxinitride coating was used to improve wear behavior and the thermal stability at higher process temperatures. Milling tests with these new AlCrN-based oxinitride coatings show an increase in the tool life primarily at a higher cutting speed (Fig. 2).

Main results of the presented solutions of titanium cutting are the choice of the proper hard coating in combination with the optimized pre- and post-treatment. For the practical use of the results a sharp cutting edge without burrs leads to the

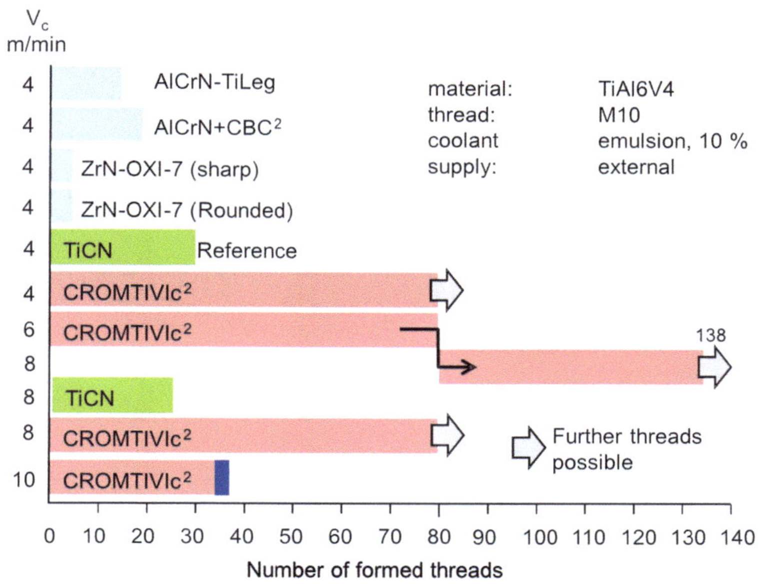

Fig. 1 Number of formed threads in Tial6V4 of different coated and prepared tapping tools

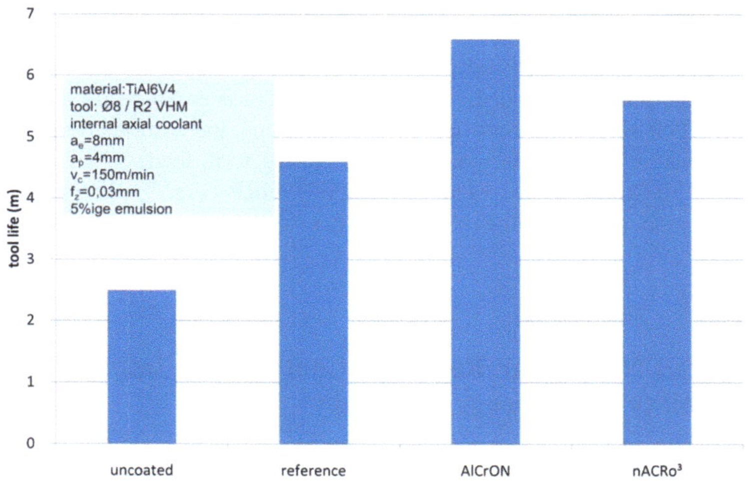

Fig. 2 Tool life of milling tools with different coatings in milling of TiAl6V4

best results in tool life. Furthermore a post treatment of the coating is necessary to remove droplets and reduce the surface roughness. The hard coating system must be chosen according to the real cutting conditions.

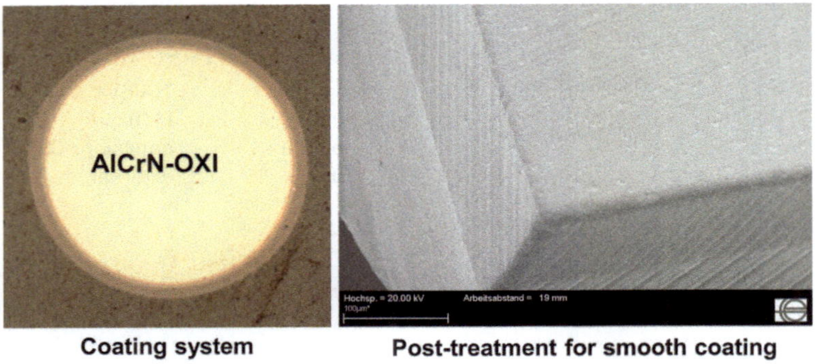

Fig. 3 Cutting of CFRP/GRP: coating system and coating post-treatment

Cutting of Fiber Reinforced Plastics

In the last years fiber reinforced plastics becomes more important due to the advantageous properties. The wide application field of these lightweight materials requires further development of technologies, processes tools and machines to evaluate and reduce delamination, fiber overlap, fibers pull out or similar phenomena. Investigations in cutting tools for CFRP/GRP are focused on the application of new and improved cutting materials, tool geometries, and new coating systems.

In cutting of fiber reinforced plastics, the application of an optimized cutting edge micro geometry and a dedicated coating system is essential for a good cut of the CFRP/GRP and a long tool life. In investigations, the combination of the new oxinitride coating with a sharp and smooth cutting tool, leads to increase in tool life up to 130 % and a sharp cut of the CFRP/GRP (Fig. 3). The cutting edge preparation and the polishing of the coating were done by the drag finishing process.

Potential of New Wear Resistant Coatings for Improved Sustainability and Energy Efficiency

Developing of sustainable innovative tools and technologies for new materials demands the application of dedicated coating systems for tools, parts and components. Depending on the used applications, new hard coating systems can be used to increase the tool life and the wear resistance as well as to increase temperature stability or to reduce friction behavior. Such new concepts are oxinitride coatings or DLC coatings as mentioned above. A relevant aspect of these new coatings is the adjustment of the optimal coating architecture such as nanocomposites, multi- or gradient layer as well as the adapted adhesion layer systems [4].

The combination of oxides and nitrides in oxinitride coatings can lead to an increase in coating properties like temperature stability or coating hardness. Figure 4 shows the higher hardness of different oxinitrides compared to nitride coatings.

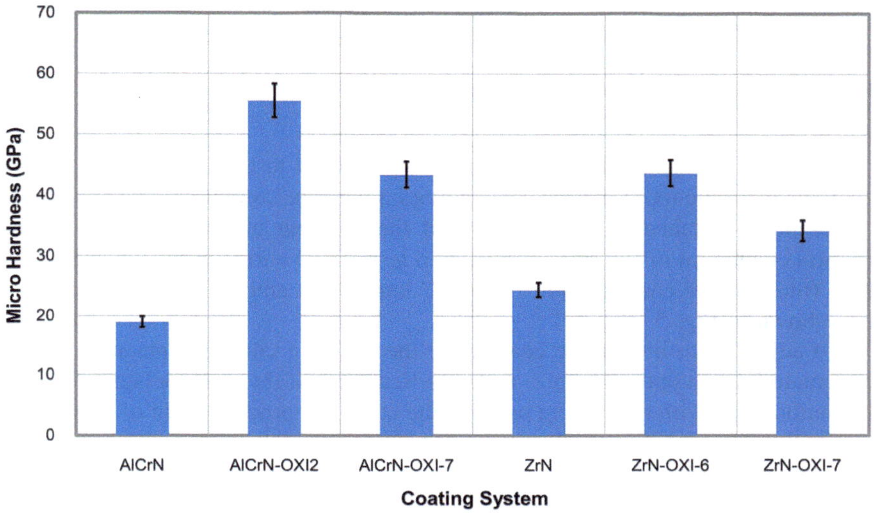

Fig. 4 Hardness of new oxinitride coating systems

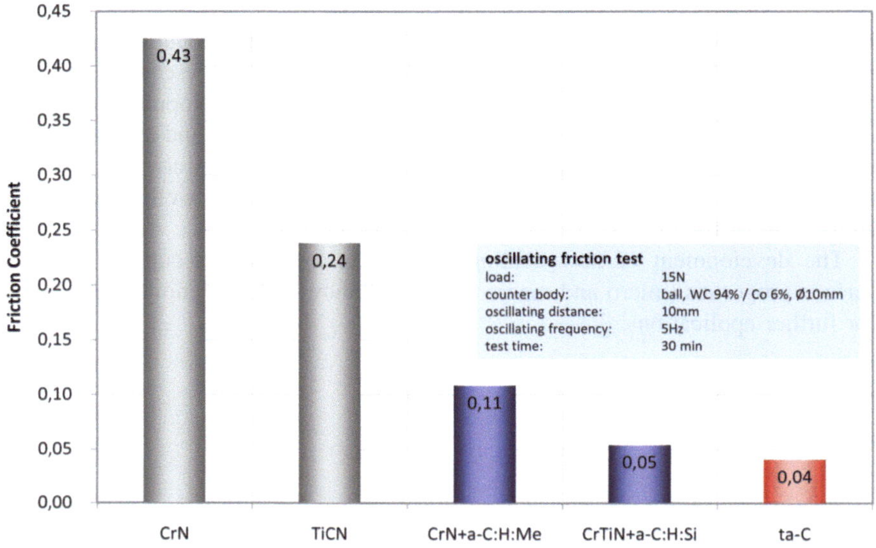

Fig. 5 Friction coefficient of different conventional and diamond like carbon coatings

Fig. 6 Necessary process chain including cutting edge preparation, coating and surface finishing

Different variations of DLC coatings can be used to obtain very low friction coefficients of the tool surfaces. Depending on the base coating, the DLC structure (amorphous, tetrahedral amorphous) and the alloying elements (metal, Si) the friction coefficient in an oscillating friction test can be varied in the range between 0.43 (nitride base coating) and 0.04 (ta-C tetrahedral amorphous carbon coating) like shown in Fig. 5.

To adjust these innovative coatings to the different cutting applications of the new materials, an optimized pre- and post-treatment of the used tools is necessary. In combination with the cutting process, the complete process chain of cutting tool production must be considered. Of special interests are the aspects of cutting edge preparation, coating deposition and surface finishing (Fig. 6).

Conclusion

In the presented contribution some issues of cutting new functional materials have been illustrated. The presented results are key aspects of the research and development of the GFE—society production engineering and development and of the industrial and research partners. Conclude the necessary tools and technologies to machine new functional materials, basic knowledge about tool construction (cutting material, geometry, coating systems) is necessary as well as the exact adjustment of the process technology on the application.

The development of new coating systems and optimized cutting edge and surface properties (micro and macro geometry) shows a high innovation potential for further applications.

References

1. Barthelmä, F.: Spanende Bearbeitung neuer Hochleistungs-Funktionswerkzeuge mit innovativen Werkzeugkonzepten. In: Die Zukunft der Fertigungstechnik, Tagungsband Wiener Produktionstechnik Kongress 2012, pp. 33–40. NWV Neuer Wissenschaftlicher Verlag, Wien (2012)

2. Gerschwiler, K., Schiffler, M.: Entwicklung von Hochleistungswerkzeugen zur Gewindeherstellung in Titanwerkstoffe; Abschlussbericht zum FKM Vorhaben Nr. 265, Heft 316 (2012)
3. Mahr, P., Frank, H., Reich, S., Barthelmä, F.: Oxygen improved hard coatings for high performance cutting processes. In: Procedia CIRP; Fifth CIRP Conference on High Performance Cutting 2012, Vol 1, pp. 208–213, (2012)
4. Barthelmä, F., Frank, H., Reich, S.: Hartstoffschichten für Werkzeuge und Bauteile – Anwendungen und Innovationen. In: Spanende Fertigung; Prozesse, Innovationen, Werkstoffe, 6. Ausgabe , S. 210-221. Vulkan Verlag, Germany (2012)

Process Force and Stability Prediction of End Mills with Unequal Helix Angles

R. Grabowski, B. Denkena and J. Köhler

Abstract The stability of cutting processes in end milling can be increased by an optimization of the tool geometry. In this paper the influence of unequal helix angles is investigated. Unequal helix angles lead to varying tooth pitches. Thus, the cutting condition of each tooth is different. A mechanistic model is presented to determine incremental process forces along the tool axis. Based on the proposed model, stability charts for tools with unequal helix angles are computed and compared with experimental investigations from literature. The presented results show that tools with an optimized geometry can increase the stability significantly.

Introduction

In metal cutting, the productivity of machining operations is often limited by chatter vibrations which occur during the cutting process. The regenerative effect, which is the main cause of chatter, is the major reason for unstable machining behavior. In order to increase the stability of end mill operations, the influence of the geometry of the tool is investigated by making various modifications, i.e., unequal tooth pitch and/or different helix angles [1]. As described by Stone, tools with unequal helix angles influence the regenerative effect [2]. Variable helix angles lead to changing tooth pitches along the tool axis. This affects the uncut chip thickness modulation, caused by the regenerative effect. With a lower uncut chip thickness modulation, the occurrence of chatter vibrations can be avoided. The method by Yusoff and Sims to predict the stability of end mills with unequal helix angles leads to a large discrepancy between experimental and predicted results [3]. With the presented method, the discrepancies can be reduced.

R. Grabowski (✉) · B. Denkena · J. Köhler
Institute of Production Engineering and Machine Tools (IFW), Leibniz Universität Hannover (LUH), An der Universität 2, 30823 Garbsen, Germany
e-mail: Grabowski@ifw.uni-hannover.de

B. Denkena (ed.), *New Production Technologies in Aerospace Industry*, Lecture Notes in Production Engineering, DOI: 10.1007/978-3-319-01964-2_13,

Mechanistic Process Force Model

Unequal tooth pitches influence the process forces. This is considered by the presented mechanistic model. Based on this model, the process stability for tools with unequal helix angles can be predicted.

Influence of Different Helix Angles on Process Forces

Figure 1 shows a section of an end mill with constant and varying tooth pitch. For a constant tooth pitch, the cross-sections of the undeformed chip A_1–A_3 are equal for all teeth. Changing the tooth pitch affects the cross-section of the undeformed chip, as shown on the right side. Due to the enlargement of the pitch p_1, the cross-section of the undeformed chip of tooth t_1 is increased. Simultaneously, the cross-section of the undeformed chip of tooth t_3 is decreased due to the smaller pitch p_3. Thus, the amplitude of the process force induced by tooth t_1 is bigger than the one induced by tooth t_3.

For a tool with different helix angles, the pitch between the teeth is changing along the tool axis. Figure 2 shows an end mill with four teeth and two different helix angles δ_1 and δ_2, whereas $\delta_2 > \delta_1$. The force vectors shown in Fig. 2 represent the maximum value of the tangential cutting force $F_{t,max,j}$ of the *j*th cutting teeth at the Process force and stability prediction of end mills with unequal helix angles 3 considered axial section z. Depending on the axial position, the tooth pitches vary, which leads to different values for the force vectors (compare **A–A** and **B–B**).

Derivation of the Process Forces

The following approach is based on the work of Lee and Altintas and is used to derive the process forces of an end mill with a constant axial immersion angle k(z) = 90° [4]. Due to the changing tooth pitches, it is necessary to integrate the

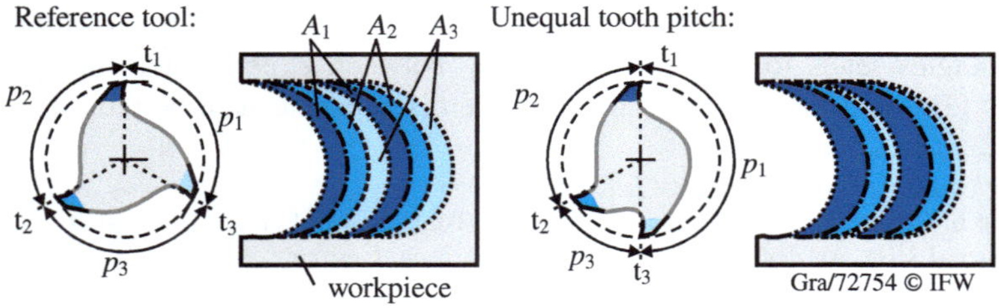

Fig. 1 Influence of tooth pitch on the cross-section of the undeformed chip

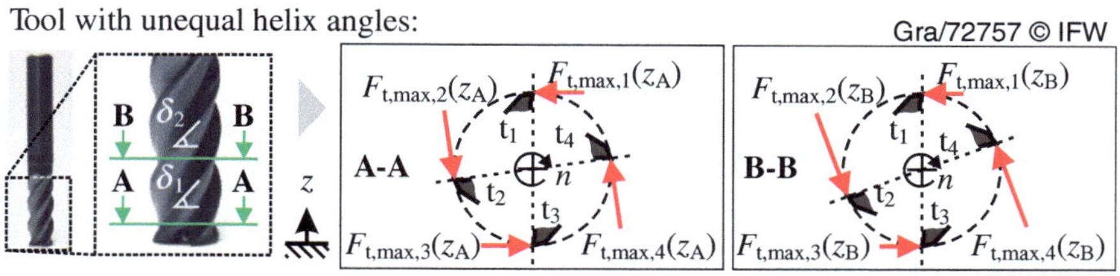

Fig. 2 Influence of the tooth pitch on process forces

tangential force $\Delta F_t(t, z_i)$ and radial force $\Delta F_r(t, z_i)$ along the z-axis to obtain the feed force $F_f(t, z_i)$ and feed normal force $F_{fN}(t, z_i)$. In the following, this integration is approximated by a sum of N_z-discrete forces:

$$\begin{bmatrix} F_f(t,z_i) \\ F_{fN}(t,z_i) \end{bmatrix} = \sum_{j=1}^{N_t} \sum_{i=1}^{N_z} \underbrace{\begin{bmatrix} \cos(-\phi_j(t,z_i)) & -\sin(-\phi_j(t,z_i)) \\ \sin(-\phi_j(t,z_i)) & \cos(-\phi_j(t,z_i)) \end{bmatrix}}_{=T(\phi_j(t,z_i))} \begin{bmatrix} \Delta F_{t,j}(t,z_i) \\ \Delta F_{r,j}(t,z_i) \end{bmatrix}. \quad (1)$$

$\mathbf{T}(\phi_j(t, z_i))$ is a rotation matrix consisting of the flute angle $\phi j(t, z_i)$ of tooth j at the ith discrete position z_i. N_t is the total number of teeth. The incremental tangential and radial forces can be obtained by the following equation:

$$\begin{bmatrix} \Delta F_{t,j}(t,z_i) \\ \Delta F_{r,j}(t,z_i) \end{bmatrix} = \left(\begin{bmatrix} K_{tc} \\ K_{rc} \end{bmatrix} h(\phi_j(t,z_i)) + \begin{bmatrix} K_{te} \\ K_{re} \end{bmatrix} \right) \Delta z g(\phi_j(t,z_i)) \quad (2)$$

with the incremental depth of cut

$$\Delta z = \frac{a_p}{N_z} \quad \text{with} \quad N_z \in \mathbb{N} \quad (3)$$

The step function $g(\phi_j(t, z_i))$ is used to determine if the current increment I of the cutting edge j is in contact with the workpiece:

$$g(\phi_j(t,z_i)) = \begin{cases} 1 & , \phi_{st} \leq \phi_j(t,z_i) \leq \phi_{eX} \\ 0 & , \text{else} \end{cases}. \quad (4)$$

ϕ_{st} is the entry angle and ϕ_{ex} the exit angle. The chip thickness $h(\phi_j(t, z_i))$ is separated into a stationary part h_{stat} $(\phi_j(t, z_i))$, based on the approximation by Martelotti [5], and a dynamic part h_{dyn} $(\theta_j(z_i))$ for stability prediction [6]:

$$h(\phi_j(t,z_i)) = \underbrace{f_i(z_i)\sin(\phi_j(t,z_i))}_{=h_{stat}(\phi_j(t,z_i))} + \underbrace{[0\;1]\mathbf{T}(\phi_j(t,z_i))^{-1} \left(\begin{bmatrix} \Delta x(t-\theta_j(z_i)) \\ \Delta y(t-\theta_j(z_i)) \end{bmatrix} - \begin{bmatrix} \Delta x(t) \\ \Delta y(t) \end{bmatrix} \right)}_{=h_{dyn}(\theta_j(z_i))} \quad (5)$$

Due to the changing tooth pitches $p_j(z_i)$, the constant feed per tooth f_z must be substituted by an effective feed $f_j(z_i)$ [7]:

$$f_j(z_i) = f_z N_t \frac{p_j(z_i)}{2\pi} \quad \text{with}\, p_j(z_i) = \begin{cases} \phi_{j-1}(t, z_i) - \phi_j(t, z_i), & j \in \{1, \ldots, z-1\} \\ 2\pi - \phi_1(t, z_i) - \phi_z(t, z_i), & j = z \end{cases} \tag{6}$$

Hence, the tangential and radial forces are also separated as follows:

$$\begin{bmatrix} \Delta F_{i,j}(t, z_i) \\ \Delta F_{r,j}(t, z_i) \end{bmatrix} = \begin{bmatrix} \Delta F_{t,\mathrm{stat,j}}(t, z_i) \\ \Delta F_{r,\mathrm{stat,j}}(t, z_i) \end{bmatrix} + \begin{bmatrix} \Delta F_{t,\mathrm{dyn,j}}(t, z_i) \\ \Delta F_{r,\mathrm{dyn,j}}(t, z_i) \end{bmatrix} \tag{7}$$

with the static part

$$\begin{bmatrix} \Delta F_{i,\mathrm{stat,j}}(t, z_i) \\ \Delta F_{r,\mathrm{stat,j}}(t, z_i) \end{bmatrix} = \left(\begin{bmatrix} K_{\mathrm{tc}} \\ K_{\mathrm{rc}} \end{bmatrix} h_{\mathrm{stat}}(\phi_j(t, z_i)) + \begin{bmatrix} K_{\mathrm{te}} \\ K_{\mathrm{re}} \end{bmatrix} \right) \Delta z g(\phi_j(t, z_i)) \tag{8}$$

and the dynamic part

$$\begin{bmatrix} \Delta F_{i,\mathrm{dyn,j}}(t, z_i) \\ \Delta F_{r,\mathrm{dyn,j}}(t, z_i) \end{bmatrix} = \begin{bmatrix} K_{\mathrm{tc}} \\ K_{\mathrm{rc}} \end{bmatrix} h_{\mathrm{dyn}}(\phi_j(z_i)) \Delta z g(\phi_j(t, z_i)) \tag{9}$$

The time delay

$$\theta_j(z_i) = \frac{p_j(z_i)}{2\pi n} \tag{10}$$

depends on spindle speed n and the tooth pitch, and thus, different helix angles influence the dynamic force components. Kleckner presented a method to predict stability charts for turning operations [8, pp. 122–124]. This method was implemented for milling tools with unequal tooth pitch by Sellmeier et al. [9]. Further extensions were made in order to take a varying tooth pitch along the z-axis, due to different helix angles, into account. These extensions are based on the presented mechanistic model. Figure 3 shows a three-dimensional diagram of the incremental stationary tangential forces $\Delta F_{t,\mathrm{stat},\, j}(t, z_i)$ for one revolution of a full-immersion operation. The end mill consists of three teeth with different helix angles and the following parameters were chosen: $K_{tc} = 794 \frac{N}{\mathrm{mm}^2}, ap = 5\,\mathrm{mm}$ and $N_z = 100$. At the bottom of the tool, the tooth pitch is the same for all teeth, which leads to $\Delta F_{t,\mathrm{stat,j}}(t, z_i) = 12$ N for all teeth. For the highest incremental axial position, the differences are apparent. For the incremental axial depth of cut Δz at $Nz = 100$, increasing and decreasing amplitudes can be seen: $N_z = 100$ ($\Delta F_{t,\mathrm{stat},1}(t, z_{100}) = 9$ N, $\Delta F_{t,\mathrm{stat},2}(t, z_{100}) = 13$ N and $\Delta F_{t,\mathrm{stat},3}(t, z_{100}) = 13.5$ N).

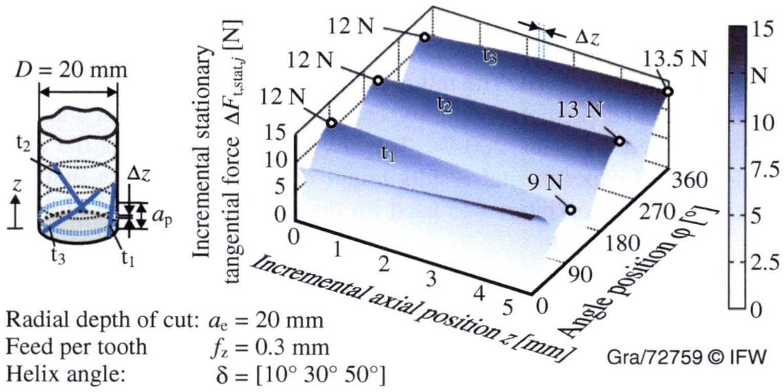

Fig. 3 Three-dimensional representation of the incremental stationary tangential forces

Experimental Validation

Yusoff and Sims investigated the influence of varying helix angles on the process stability [3]. Their experimental data is used to evaluate the proposed method. To predict stability charts, they approximated the cutting process by a one degree of freedom model. The modal parameters are used for the prediction of stability charts. Figure 4 shows the experimental and predicted stability charts. On the left diagram, the investigated tool had a constant tooth pitch and helix angle. The experimentally determined minimum stability limit for all spindle speeds n can be found at an axial depth of cut $a_{p,\ min} = 0.3$ mm, which agrees well with the

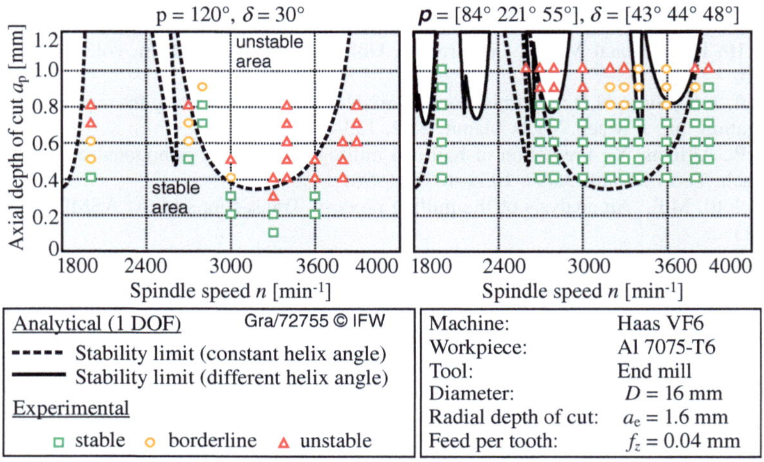

Fig. 4 Stability charts for an end mill with constant and different helix angles (experimental data from [3])

calculated stability chart. The right diagram shows the results for a tool with unequal tooth pitch and different helix angles. In comparison to the reference tool, the stability limit is increased. The minimum stability limit is $a_{p,\,min} = 0.7$ mm for the experimental and predicted stability chart. Thus, the minimum stability limit could be increased by over 200 %. The discrepancies between the analytical stability lobes and experimental results might result from the fact that only one mode of the flexible workpiece was used for the mechanistic model of the milling process.

Conclusions

Tools with optimized geometric shapes, e.g., different helix angles, can improve the process stability, and thus, the productivity. For a solid validation of the presented method, further experiments must be carried out. This also includes modeling a milling system with multiple eigenmodes and an in-depth analysis of the experimental and predicted chatter frequencies.

Acknowledgments This work has been supported by the Ministry for Science and Culture of Lower Saxony (MWK) within the excellence cluster 'Pro3gression'.

References

1. Sims, N., Mann, B., Huyanan, S.: Analytical prediction of chatter stability for variable pitch and variable helix milling tools. J. Sound Vib. **317**(3–5), 664–686 (2008)
2. Stone, B.: The effect on the chatter behavior of machine tools of cutters with different helix angles on adjacent teeth. In: Advances in Machine Tool Design and Research (Proceedings of the 11th International MTDR Conference), University of Birmingham, vol. A, pp. 169–180 (1970)
3. Yusoff, A.R., Sims, N.D.: Optimisation of variable helix tool geometry for regenerative chatter mitigation. Int. J. Mach. Tools Manuf. **51/2,** 133–141 (2010)
4. Lee, P., Altintas, Y.: Prediction of ball-end milling forces from orthogonal cutting data. Int. J. Mach. Tools Manuf. **36**(9), 1059–1072 (1996)
5. Martelotti, M.E.: An analysis of the milling process. Transactions of the ASME **63**, 667–695 (1941)
6. Sellmeier, V., Denkena, B.: Stable islands in the stability chart of milling processes due to unequal tooth pitch. Int. J. Mach. Tools Manuf. **51**(2), 152–164 (2011)
7. Engin, S., Altintas, Y.: Generalized modeling of milling mechanics and dynamics: part 1: helical end mills. Am. Soc. Mech. Eng. Manuf. Eng. Div. (MED) **10**, 345–352 (1999)
8. Kleckner, J.: Ein Beitrag zur Analyse dynamischer Interaktionen bei selbsterregungsfShigen Drehbearbeitungsprozessen. PhD thesis, TU Darmstadt, August 2001
9. Sellmeier, V., Denkena, B., de Leon, L.: Impact of the tooth pitch on the process stability of milling. In: Proceedings of the 1st International Conference of Process Machine Interactions (2008)

High Rate Production of Laminar Wing Covers With Modular "Shoe Box" Tooling

Markus Kleineberg and Matthias Grote

Introduction

Lightweight design is one of the major drivers for primary airframe structures because efficiency of future aircrafts is the dominating sales argument. CFRP (**C**arbon **F**iber **R**einforced **P**lastic) based airframe components do have the potential to challenge sophisticated aluminum structures but the complexity of the production sequence, the number of production parameters that have to be optimized and last not least the natural tolerances of the highly developed raw materials are crucial challenges for the development phase. In 1983 Airbus presented the A310 with an innovative composite VTP (**V**ertical **T**ail **P**lane) and by doing this Airbus set the course for today's fully composite based aircrafts like the Boeing 787 and the Airbus A350XWB (Fig. 1). Drag reduction is another important enabler for future aircraft efficiency. To achieve this drag reduction it is a primary goal to produce wing structure with extremely smooth and precise surfaces in order to provide laminar flow capabilities at high Mach numbers.

Design for Production Aspects for Composite Components

Even though various successful composite design and production concepts have been developed in the last 30 years it is still worth looking at the principle idea of the very first composite VTP of the Airbus A310. In contrast to many differential

M. Kleineberg (✉) · M. Grote
Deutsches Zentrum für Luft- und Raumfahrt e.V. (DLR), Institut für Faserverbundleichtbau und Adaptronik, Faserverbundtechnologie, Lilienthalplatz 7, 38108 Braunschweig, Germany
e-mail: markus.kleineberg@dlr.de

B. Denkena (ed.), *New Production Technologies in Aerospace Industry*,
Lecture Notes in Production Engineering, DOI: 10.1007/978-3-319-01964-2_14,

Fig. 1 Boeing 787 with composite airframe

composite solutions (joined skin, stringer and rib components) that could be characterized as "Black Metal" the highly integral A310 VTP design (skin with integrated stringers and rib posts) was consequently tailored to utilize the CFRP advantages and minimize joining operations. The resulting structural component was highly competitive from a weight efficiency point of view but the manufacturing approach was optimized for a very low production rate of only 1–5 aircrafts per month with a very high amount of manual interaction during manufacturing. Today's single aisle aircrafts like the Boeing 737 or the Airbus A320 are produced with a rate of 40 and more aircrafts per month which means that the manufacturing strategy has to ensure a production window of less than 8 h for one aircraft.

In order to limit the overall production risk for composite structures there still is a tendency to produce large but rather simple "Black Metal" structures with reduced structural integration and accept the resulting increased assembly effort and weight penalty. What might have been underestimated in this respect is that the geometrical precision of composite structures is heavily influenced by various specific effects that have to be dedicated to the anisotropic nature of CFRP. Especially laminate thickness variation due to variable fiber content and "Spring-In" deformations at curved sections lead to increased scatter in geometrical accuracy and subsequently to expensive and time consuming rework and shimming operations during assembly.

Fig. 2 Fully integral "Shoe Box" wing skin approach

The "Shoe Box" Wing Skin Design

Taking the above mentioned experiences into account the selected approach for a future high rate production composite wing cover with laminar flow capabilities in the ongoing BMWI (**B**undes**m**inisterium für **Wi**rtschaft und Technologie) founded LUFO (**Lu**ftfahrt**fo**rschungsprogramm) program was to go back to the very roots of composite applications and realize a highly integral design with selected modifications to ensure an efficient production and assembly process. The major prerequisites were maximized geometrical accuracy in all joining areas and minimized cycle times. The high degree of structural integration in the backing structure provides sufficient global rigidity to minimize process induced deformations on the aerodynamic outer surface which in turn ensures laminar flow capabilities. In order to simplify manufacturing all undercuts have been eliminated e.g. by parallelizing stringer blades and by choosing a T-Stringer design (Fig. 2).

The "Shoe Box" Manufacturing Approach

A major prerequisite for the "Shoe Box" manufacturing approach which has been developed in close cooperation with Airbus und Premium Aerotec was that existing process certification rules can be applied. Even though the final wing cover structure is fully integrated and based on compatible prepreg semifinished products it has been decided to choose different preparation concepts for the flat skin laminate and the bi-directional backing structure.

In case of the flat skin laminate a robot based multi head ATL (**A**utomatic **T**ape **L**aying) concept has been chosen because this approach allows to increase the lay-up rate without exceeding a critical lay-up speed and limits the investment costs because highly flexible standard robots and rail systems can be used (Fig. 3).

To realize the bi-directional backing structure it has been decided to wrap box like tooling modules with prepreg fabric (Fig. 4) and combine the tooling modules in a way that stringers and rib caps are formed. To provide the required

Fig. 3 DLR GroFi production center in Stade

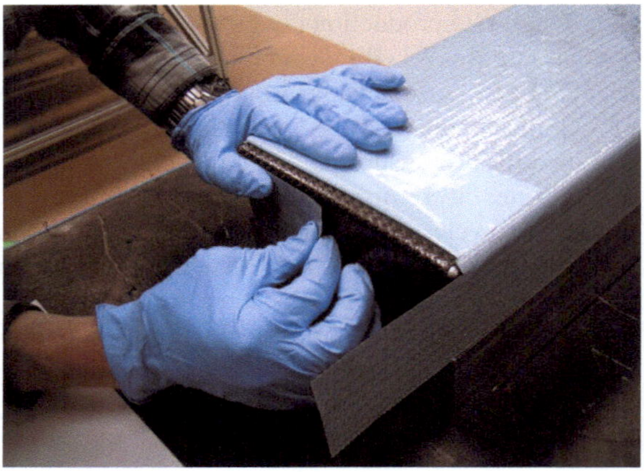
Fig. 4 Wrapping of tooling modules

consolidation pressure for the prepreg a material mix of aluminum and Invar is used where the thermal expansion of the aluminum compresses the prepreg stack.

To provide the required global geometrical precision each tooling module is directly linked to an Invar base plate. In order to provide maximum precision in the joining area the fixed bearing of the tooling module is located at the surface that forms the joining area of the rib cap. Since it is a known problem that heating and cooling of massive tooling modules is critical for the cycle time hollow tooling modules where applied. To further increase heat-up and cool down rates the airflow of the autoclave was lead through the tooling modules (Fig. 5).

Skin lay-up and backing structure lay-up together with their dedicated tools are then combined and the resulting tooling can be described as a closed autoclave mould without the typical vacuum bag of classic open moulds (Fig. 6).

A further innovation that shall be demonstrated with the modular "Shoe Box" tooling is a quality assurance and logistic feature, integrated in each tooling

Fig. 5 Hollow aluminum tooling module

module to monitor the process chain. With respect to the industrial production scenario this is an approach to establish a robust and self-adjusting wing production environment where each tooling module knows its current status from a wear point of view, its position in the facility and all other quality relevant process data like e.g. the dedicated lay-up sequences and flat patterns. Special high temperature and resin resistant RFID chips have been identified to enable this feature.

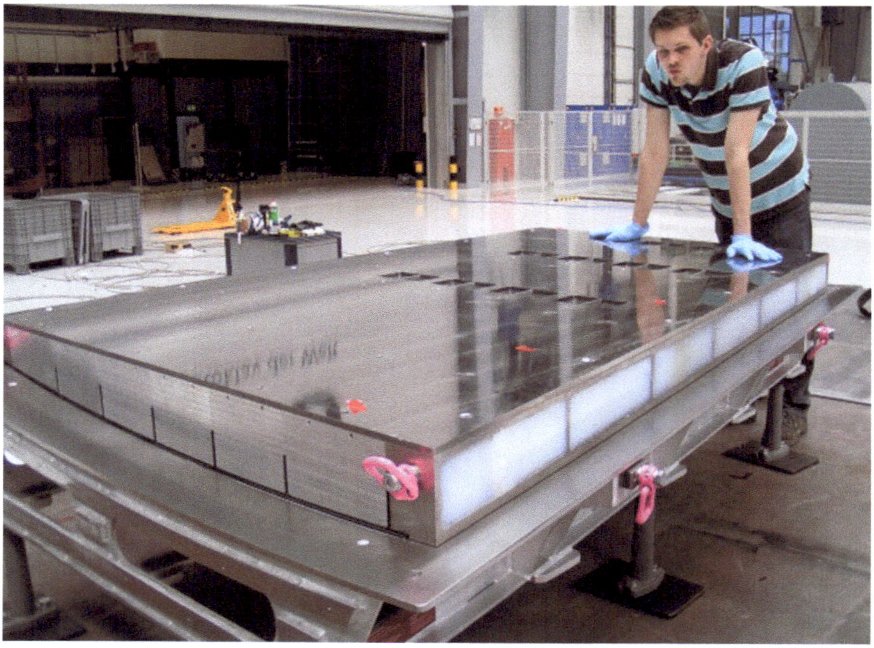

Fig. 6 Final tooling setup

For future applications it is also planned to integrate specialized sensors and actuators in the tooling modules in order to be able to interactively adjust the processing conditions (e.g. airflow through the core) to what is exactly required in the on-going process. Process compatible Piezoceramic elements and the dedicated analyses hardware have been developed to monitor all relevant process parameters like e.g. degree of cure and gaps between adjacent tooling components.

A constant data exchange shall ensure, that all relevant production parameters are within their limits and that all logistic action can automatically be adjusted to the current process status. Furthermore each "Shoe Box" tooling module shall have the ability to optimize its local thermal boundary conditions by actively adjusting the airflow through the hollow core. Up to a certain level variations in the cure behaviour of the resin system and deviating fiber contents in the semi-finished product shall also be detected and compensated by adapting the process parameters.

Fig. 7 "Shoe Box" validation structure and *gom Atos* measurement system

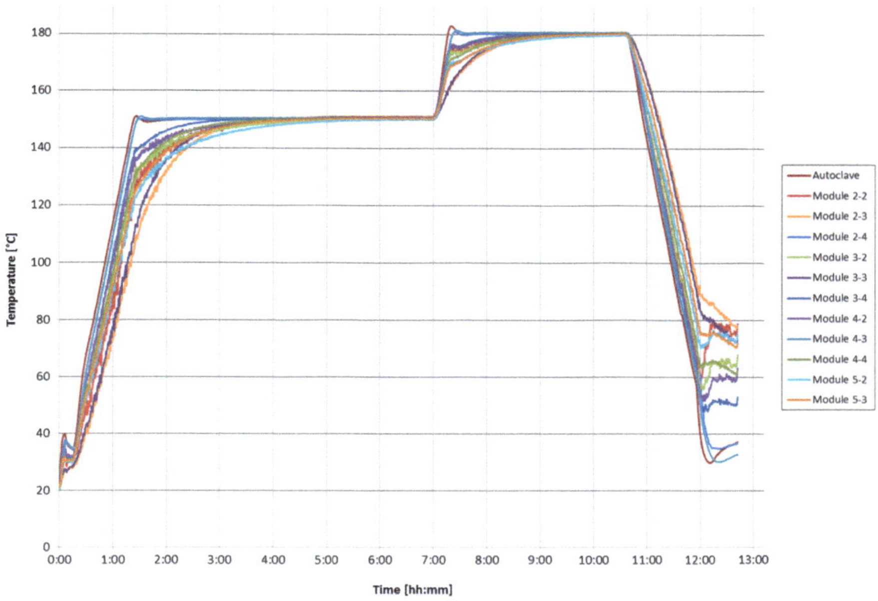

Fig. 8 Tooling module and autoclave temperature

Concept Validation

To validate the "Shoe Box" concept a first manufacturing trial under representative boundary conditions has been carried out. To prove the technology readiness status a typical swept angle, tapering and a skin curvature of an outer wing section has been chosen for the large scale manufacturing trial.

For the first manufacturing trial highest priority was given to the approval of the basic hollow core "Shoe Box" strategy which means that thermal behavior of selected representative tooling modules was analyzed in detail (Fig. 8).

The temperature plot shows that the tooling module temperature is only slightly lagging behind the autoclave temperature even though the airflow through the module may still be further improved.

Compression of the prepreg in the skin and the backing structure laminate was sufficient to ensure the required fiber content and also precision of joining areas is within the addressed target. To ensure the geometrical accuracy for NLF (**N**atural **L**aminar **F**low) wings, each part was 3D measured and analyzed using the *gom Atos* system (Figs. 7 and 9).

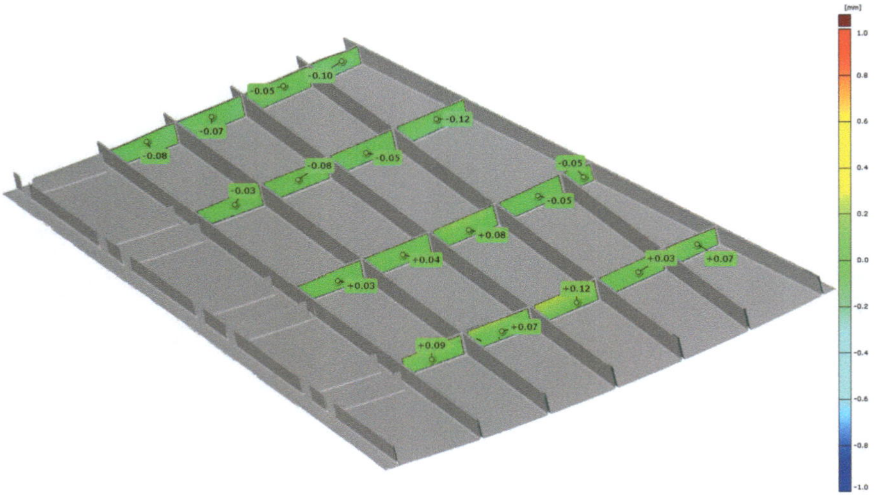

Fig. 9 Geometrical accuracy of joining areas

Conclusion

The basic strategy to produce large, high precision wing cover components with integrated stiffeners in a production window of 8 h has been validated with respect to autoclave cycle time and geometrical component quality. Other aspects like the addressed multi head ATL lay-up procedure, the online quality assurance and process control features and the RFID based logistic concept are still under development and will be validated in one of the following manufacturing trials.

The described manufacturing approach shows that it is possible to further extend productivity of composite production scenarios even under rather conventional and certified manufacturing boundary condition.

Simulation of Residual Stress Related Part Distortion

Berend Denkena and Steven Dreier

Abstract During the machining of structural components, material inherent stresses are removed and additional residual stresses are induced into the boundary layer of the workpiece. Both effects can result in distortions of the workpiece which leads to shape deviations and time consuming repair processes or even scrap. The research objective of this paper is the simulation of distortion caused by residual stress for machined aircraft components to avoid an insufficient quality. The method, presented in this paper, is able to predict the part distortion caused by process induced and initial residual stresses. A custom FEM pre-processing is used which takes the actual machining strategy into account to predict the shape deviations of the machined workpiece. An experimental verification is given by comparing measured part distortions caused by a manufacturing process and predicted by the developed simulation. The Validation is done on a complex structural part. The simulation results show good agreement between simulated and measured part distortions.

Keywords Distortion simulation · Structural components · Process induced residual stresses · Finite element method

Introduction

The aircraft manufacturer Airbus forecasts that the worldwide fleet of civil passenger and cargo aircraft will increase from 17,171 in 2011 to 35,489 in 2031 [1]. The demand for new aircrafts raises the challenge to increase productivity and quality in all production stage. Modern aircrafts are designed in order to maximize

B. Denkena · S. Dreier (✉)
Institute of Production Engineering and Machine Tools (IFW), Leibniz Universität Hannover, An der Universität 2, 30823 Garbsen, Germany
e-mail: dreier@ifw.uni-hannover.de

B. Denkena (ed.), *New Production Technologies in Aerospace Industry*, Lecture Notes in Production Engineering, DOI: 10.1007/978-3-319-01964-2_15,

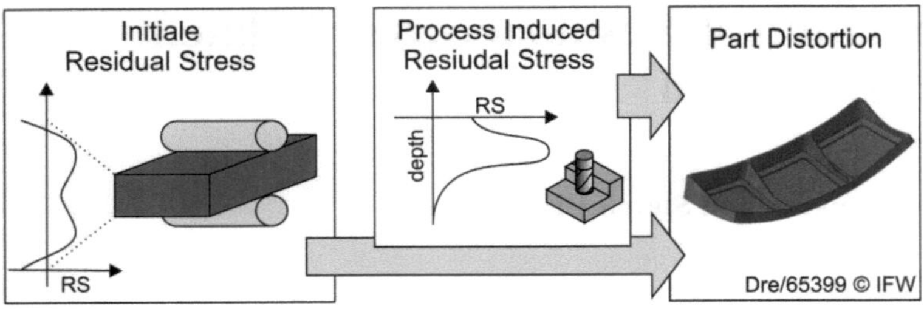

Fig. 1 Causes for part distortions due residual stress (RS)

energy efficiency. This denotes that low fuel consumption is in focus of the development of new aircrafts such as the Boeing 787 or the Airbus A320neo [2]. A decisive factor for the fuel consumption is the weight of the aircraft. In order to maintain low weight, the supporting structure components are designed in a way that a maximum rigidity is achieved with minimal residual material thickness. This results in parts with a typical thickness close to 2 mm. Such components can be up to 14 m long and additionally show complex geometries. In addition, the chosen materials are difficult to machine and result in several problems for the machining process. Especially, part distortions occurring after machining cause time consuming repair processes or even scrap. During the machining process of structural components, up to 90 % of the material is removed from the blank. Thereby, material inherent residual stresses are removed and additional residual stresses are induced into the boundary layer of the machined workpiece by the machining process. Both effects cause distortions of the workpiece and lead to shape deviations which cannot be repaired without damaging the aircraft component (Fig. 1). To ensure a more efficient production process, these phenomena have to be predicted and avoided in advance.

State of the Art: Process Induced Residual Stresses and Distortions

Residual stresses can be attributed to mechanical and thermal loads, which occur during machining process in an independent manner [3]. Various research projects have so far dealt with the problems of process induced residual stresses for cutting processes. Tönshoff analysed the influence of the cutting process on the generation of residual stresses in the boundary layer of the machined workpiece. He found out, that the process induced residual stresses depend on the chosen cutting conditions [4]. Norihiko et al. examined the influence of cutting speed on residual stresses for the machining of Ti6Al4V using diamond tools [5]. Klocke examined the rough machining of Ti-6242 alloy, Ti-6246 and Udimet720Li, looking at the vibration strength, the hardness profile and the residual stresses of these titanium and nickel-based alloys. Thereby, it could be shown that occurring residual stresses are material dependent [6]. Research shows that the process induced residual stresses strongly depend on the process conditions. Factors of influence are the chosen tool geometry and the material as well as the cutting and cooling conditions [7, 8].

To be able to calculate the influence of residual stresses on the part distortion, the global distribution of stresses in the workpiece has to be taken into account during the machining process. By measuring the initial stress distribution and including it in a finite element model the distortion of the machined workpiece can be modelled [9]. Ratchev and Afazov predicted residual stress for turning of Ti6Al4V and mapped them into an axisymmetric FE model to predict distortions of turned parts [10]. Brinksmeier et al. developed a new method to predict the shape deviation of machined workpieces with complex geometry with a finite element simulations [11]. Marusich et al. reported on a finite element method software for the prediction of part distortion. The distortion of the machined workpiece is calculated by a simplified shell model of the workpiece [12]. Zhongyi analyzed the effect of residual stresses on machining distortion on an aircraft structure part. He used an FE model with static element deactivation technique to simulate the cutting process. Furthermore he used thermal–mechanical coupling analysis to predict initial residual stresses [13]. Bi et al. predicted the distortion of aerospace monolithic components. They considered initial residual stress, cutting loads, fixture layout, cutting sequence, and tool path to build a FE model for the prediction of part distortions caused by a milling process [14].

In order to avoid part distortion in machining of complex aircraft components, the accuracy of the machining simulation must be increased. To validate the influence of different machining strategies on the phenomenon of part distortion, the simulation has to consider the tool path. Additionally, the influence of initial residual stresses, e.g. from the previous manufacturing step has to be included into the simulation [15].

Simulation of Process Induced Part Distortion

Within the process planning phase the manufacturing process is designed to achieve an optimal workpiece quality. If quality failures occur, they will be identified first during the quality inspection. In order to avoid quality inaccuracies in advance, a process simulation for the prediction of process induced distortions was developed. Based in a simulation of the machining process, induced distortions can be calculated and avoided by the adaption of the machining strategy or the process conditions. The developed finite element simulation is based on information from the process planning stage. Based on the workpiece and process data from the CAD/CAM system the workpiece is prepared for the simulation. The tool path is used to identify machined areas and to implement the loads from the process induced residual stresses. Finally, the FE-model can be solved by a

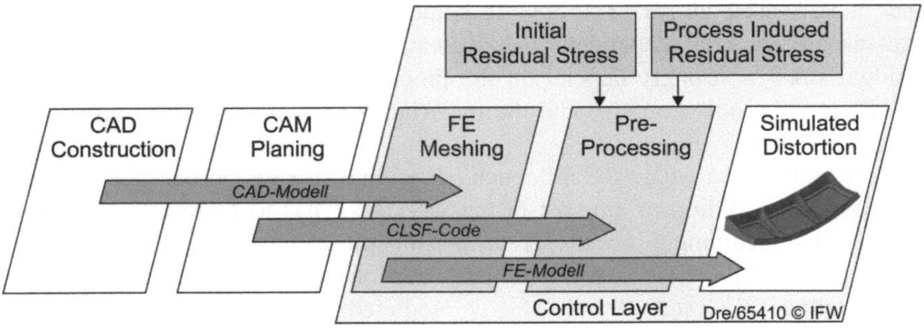

Fig. 2 Information structure of the developed distortion simulation

commercial finite element analysis (FEA) system, delivering detailed information about the distortion of the workpiece geometry. The subsequent Fig. 2 shows the information structure of the developed distortion simulation:

The core of the simulation structure is a control layer which transfers the simulation objects to the several simulation steps.

FE Meshing

In the first step, the workpiece geometry of the final component is transferred to the meshing process, using standardized CAD formats. The meshing of the workpiece and the pre-processing is controlled by a script. For the meshing of complex geometries with variable element size a volumetric tetrahedron element is chosen. This element ensures an automated generation of accurate meshes for complex geometries in most cases. Regarding memory consumption the initial mesh is coarse with a typical element size of 2 mm. In order to implement process induced loads into the machined surfaces of the workpiece the resolution of the FE mesh in the boundary layer has to be increased in the following step.

Pre-processing

In order to simulate the effect of machining induced distortions due to residual stresses, the tool movement and the engagement conditions of the cutting process have to be determined in the second step. This is done by a process simulation, which uses the tool path and the process parameters provided by a Cutter-Location-Source-File (CLSF) from the CAM software. The process simulation determines the machined boundary elements of the workpiece mesh for the refinement of the boundary layer. The developed algorithm for the refinement of the FE mesh on the surface layer is based on a sweep volume of the tool path (Fig. 3).

The refinement of the surface mesh works by the orthogonal displacement of node positions and the squeezing and generation of elements, as it can be seen in Fig. 3. The displacement is inverse to the surface normal N.

After the mesh refinement in the boundary layer, process induced residual stresses are obtained from an empirical data base and included into the FE model. Therefore, the directions of the tool axis and feed are determined for the transformation of residual stresses to the coordinate system of the boundary mesh of the machined workpiece. Figure 4 shows a typical residual stress distribution in the boundary layer caused by end milling operations of aluminum.

The representation of residual stress profiles in the finite element model is done by the discretization of the boundary layer into several layers with constant residual stress. To investigate the optimal discretization of boundary layers, an analogy was used by taking a simplified beam model with different numbers of discretization layers into account. It was determined that a representation of residual stress profiles with only one layer is not sufficient. The distortion results differ from the results with 10 regular layers by 14.4 %. Further analysis showed, that representing the curve with two layers of irregular depth divided at the curve minima, as it is shown in Fig. 4 right, reduces the error to 0.01 % compared to 10 regular layers. This representation minimizes the number of elements for the

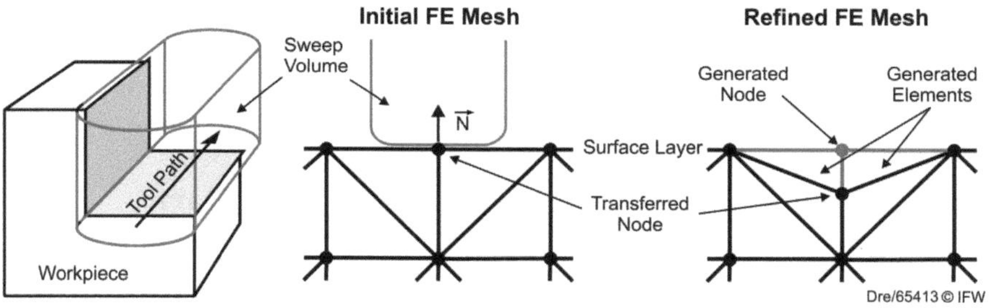

Fig. 3 Algorithm for FE-mesh refinement

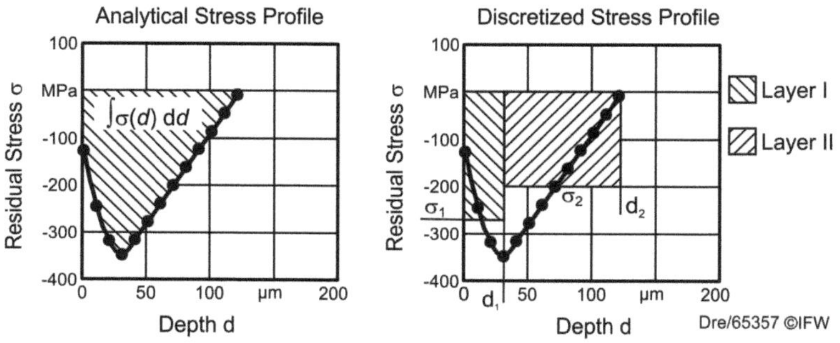

Fig. 4 Residual stress distribution represented by boundary layers

refined mesh. The implementation of stress profiles is done by the control script, which links the local residual stress tensors to the refined mesh elements. These tensors consist of the initial and process induced residual stresses. The resulting process induced residual stress profile is determined by X-ray diffractometry. The stress value for the two boundary layers is calculated by integrating the affected part of the curve (Fig. 4). The stress tensors are transformed into the workpiece coordinate system according to the feed direction and the tool axis which were computed in the process simulation.

Solving

In this step of the simulation, the completed FE model is transferred to the FE solver. The simulation structure enables the communication to arbitrary FEA software. For common FEA software initial residual stresses can be defined via stress tensors for several elements. By providing the stresses for all elements and solving the resulting finite element model by a linear static analysis, it is possible to calculate the distortion of the workpiece resulting from the implemented loads. By the superposition of initial residual stresses from previous processing steps and process induced stresses from the current machining process, combined distortion effects can be determined. The simulation results can be examined by common post processors by rendering the simulation data for the deflection of the machined workpiece.

Validation of the Simulation of Process Induced Part Distortion

In order to verify the developed simulation approach against real part distortions, an exemplary analogy part was designed based on an aircraft structural component of the Airbus A240 XWB (Fig. 5). The selected workpiece material was aluminum alloy AN EW 7075 in low-stress condition T651. The analogy part was manufactured from rolled plates with dimensions 70 × 330 × 480 mm. The initial residual stress state of the plate material was determined by a reference sample with the X-ray diffraction method of a reference sample. During manufacturing the part is clamped so no machining related distortion could occur. After the milling process the part was unclamped and relaxed to its distorted state.

For the production of the analogy part two tools were used. One is a double-edged end mill with a diameter of D = 20 mm without corner radius (tool 1). This tool was used for the planar bottom side of the part. From preliminary experiments, it is known that this tool induces is high residual stress into the machined surface. The top side of the component with the pocket structures was milled with a four-

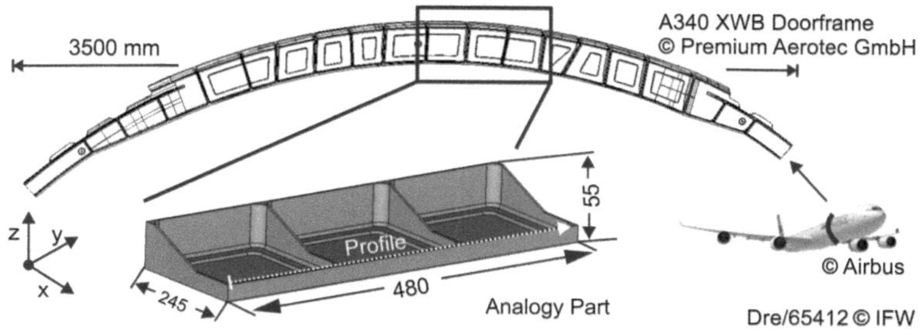

Fig. 5 Workpiece based on structural component

edged D = 20 mm end mill with corner radius of $r_\varepsilon = 4$ mm (tool 2). This tool induces low residual stress. For both tools the residual stress depth profiles were determined from reference samples with the X-ray diffraction method. The residual stress profile for tool 1 was discretised with two layers ($d_1 = 51$ µm; $\sigma_{1\|} = -174$ N/mm^2; $\sigma_{1\perp} = -150$ N/mm^2/$d_2 = 250$ µm; $\sigma_{2\|} = -93$ N/mm^2; $\sigma_{2\perp} = -57$ N/mm^2). Due to the low residual stress tool 2 was discretised with one layer ($d_1 = 100$ µm; $\sigma_{1\|} = -21$ N/mm^2; $\sigma_{1\perp} = -20$ N/mm^2).

Based on the workpiece geometry, the tool path and the cutting conditions the distortion of the workpiece was determined by the developed simulation approach. The generated volumetric finite element model consisted of 174.871 nodes and 750.512 elements. An elastic material model with a Young's Modulus of 72 GPa and a Poisson's Ratio of 0.34 was used. To simulate a free distortion no boundary conditions were applied.

The machined and the simulated workpiece was measured on a line in Y direction (Fig. 5, profile). The resulting profile can be found in Fig. 6. In the referenced CAD model this profile is a straight edge.

The process induced residual stress led to a significant part distortion of the machined part with a maximum of 0.510 mm on the measured profile. The comparison of machined and simulated distortions can be seen in Fig. 6. On the path of the measured profile the predicted distortion matches the measured distortion very accurate. The maximum distortion, computed by the FEM simulation, is 0.469 mm leaving an error of 0.051 mm (9.81 %). Despite small

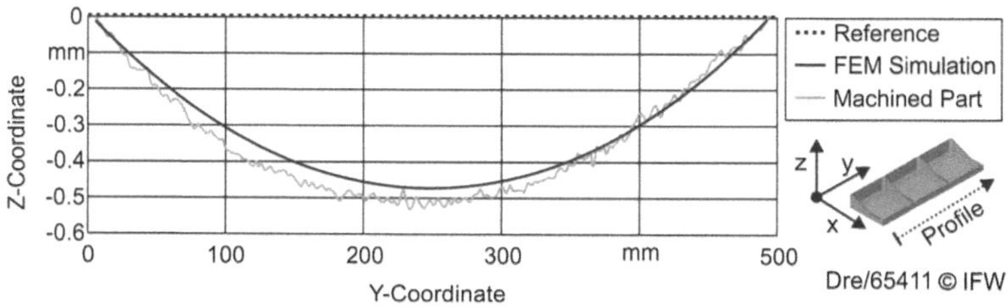

Fig. 6 Comparison of machined and simulated distortion results

deviation the developed simulation approach can be considered as proven for the given test case.

Summary and Outlook

The influence of initial and process induced residual stress distribution cause significant part distortions after the machining of large structural components. The simulation approach, presented in this paper, was developed to predict process induced part distortions according to the part geometry, the cutting conditions and the machining strategy. This provides possibilities to avoid or minimize part distortions in advance by the adaption of machining strategies or process conditions.

The presented simulation uses the finite element method with a volumetric tetrahedral mesh to calculate the resulting distortion of the workpiece. To represent the thin residual stress layers a remeshing algorithm was developed to refine the mesh structure underneath the machined surfaces. By this detailed simulation approach it is possible to analyse the impact of local machining strategies on the global distortion of large components. As demonstrated in Chap. 4, the developed simulation structure achieves sufficiently precise predictions for aluminum components using process induced residual stress measured by the X-ray diffraction method.

Further research objectives will be the development of a scalable simulation structure for the distortion simulation of large aircraft integral components with lengths higher than 1 m. By analyzing the impact of different strategies and process parameters on the overall part distortion a general guideline for an optimal process layout in terms of minimized distortions will be developed. Furthermore, the application of the developed simulation approach into industrial process planning will reduce cost-intensive repair processes.

Acknowledgments This work has been funded by the Ministry of Economics, Labour and Transport of Lower Saxony within the framework of research and technology projects for the aviation industry of Niedersachsen through the project *QualiTi: Method for the modeling of effects on the workpiece quality during the milling of titanium.*

References

1. Airbus GMF 2012: Airbus Global Market Forecast 2012–2031, Toulouse, Dec 2012
2. Deutsche Lufthansa AG: Nachhaltigkeitsbericht Balance, Frankfurt am Main, Mai 2012
3. Brinksmeier, E., Cammett, J.T., König, W., Leskovar, P., Peters, J., Tönshoff, H.K.: Residual stress—measurement and causes in machining processes. Ann. CIRP **31**(2), 491–510 (1982)
4. Tönshoff, H.K.: Eigenspannungen und plastische Verformungen im Werkstück durch spanende Bearbeitung. Dr.-Ing. Diss., Techn. Hochschule Hannover, Garbsen (1966)

5. Norihiko, N., Akio, M., Suquru, M.: Study on machining of titanium alloys. CIRP Ann. **32**, 65–69 (1983)
6. Klocke, F.: Bearbeiten von Titan- und Nickelbasislegierungen. Einfluss der spanenden Bearbeitung auf die Bauteileigenschaften. VDI-Z Integrierte Produktion **149**(3), 70–72 (2007)
7. Tönshoff, H.K., Denkena, B.: Spanen – Grundlagen. Springer, Berlin (2011)
8. de Leon, L.: Residual stress and part distortion in milled aerospace aluminium. Dr.-Ing. Diss., Leibniz University Hannover, Garbsen (2010)
9. Hornbach, D., Prevey, P.: Development of machining procedures to minimize distortion during manufacture. In: Proceedings of the 17th Heat Treating Society Conference and Exposition, pp. 13–18, ASM, Metals Park, Ohio, 1998
10. Ratchev, S.M., Afazov, S.M., Becker, A.A.: Mathematical modelling and integration of micro-scale residual stresses into axisymmetric FE models of Ti6Al4V alloy in turning. CIRP J. Manuf. Sci. Technol. **4**, 80–89 (2011)
11. Brinksmeier, E., Sölter, J.: Prediction of shape deviations in machining. CIRP Ann. **58**, 507–510 (2009)
12. Marusich, T.D., Usui, S., Terauds, K., Becker, B.V.: Finite element modelling of part distortion. In: Proceedings of the 2nd International Conference in Distortion Engineering, pp. 133–142, Bremen, 17–19 Sept 2008
13. Zhongyi, M., Yunqiao, W.: Analyzing distortion of aircraft structural part in NC machining based on FEM simulation. In: International Conference on Mechanical and Electrical Technology ICMET (2010)
14. Bi, Y., et al.: Machining distortion prediction of aerospace monolithic components. J. Zhejiang Univ. Sci. A **10**, 661–668 (2009)
15. Denkena, B. (Hrsg.): Neue Fertigungstechnologien in der Luft- und Raumfahrt. Begleitband zum Seminar, Hannover, 24. und 25. November 2009, Berichte aus dem IFW, Band 06/2009, PZH Produktionstechnisches Zentrum GmbH

Increasing Accuracy of Industrial Robots in Machining of Carbon Fiber Reinforced Plastics

Martin Freising, Simon Kothe, Markus Rott, Hendrik Susemihl and Wolfgang Hintze

Abstract Compared to conventional milling machines industrial robots (IR) using serial kinematics bear great potential concerning cost-efficient and productive machining. However, the main disadvantage is their poor accuracy, caused by low structural stiffness and gear backlash. A novel robust approach for accuracy optimization of IR in edge milling is presented. This is highly relevant for manufacturing structural components made of carbon fiber reinforced plastics (CFRP). As a result of the long lever between the tool center point and the first rotational axis, backlash of the first gear mainly contributes to the overall machining accuracy of the IR. In order to eliminate this, the first axis is preloaded by a torque, caused by the gravitational force of the attached robot structure. Therefore, the IR is mounted on an inclined base. The performance of this setup is experimentally evaluated. Improvements of up to 50 % in terms of static stiffness, deviation of circular movement and contour accuracy in edge milling have been achieved.

Introduction

The increasing use of carbon fiber reinforced plastics (CFRP) in modern aircrafts [1] requires productive and cost-efficient manufacturing technologies for large shell structures. In order to fulfill the high precision demands, such parts need to be machined subsequent to the autoclave process itself. Today this is accomplished mainly by heavy large-volume machining centers. However, these are expensive and suffer from low productivity. Therefore, industrial robots (IR) can be a promising alternative. Although IR are well-established in aerospace industry for handling automated drilling and riveting units [2], their application for milling

M. Freising (✉) · S. Kothe · M. Rott · H. Susemihl · W. Hintze
Hamburg, Germany

B. Denkena (ed.), *New Production Technologies in Aerospace Industry*,
Lecture Notes in Production Engineering, DOI: 10.1007/978-3-319-01964-2_16,

processes is still a challenge [3], due to the required path accuracy with simultaneously occurring process forces [4].

As a precondition for a flexible use of IR in an automated CAD/CAM process chain, their absolute position accuracy must be significantly improved. Suitable methods reaching the necessary level of calibration have been developed [5]. Several authors studied the elastic deformation of IR related to the cutting force in milling. Online control systems and offline models for compensating such deflections have been developed recently [4, 6–8]. Control concepts on the output side of the gears as well as modifications of the robot structure are possible solutions to eliminate loss in accuracy due to backlash [9]. Additional force control systems supporting these techniques are available on the market [10]. These approaches are intricate and therefore contradict the concept of a low cost machining solution using IR.

Concept of Inclined Mounted Robot

The gear of the first rotational axis of serial IR kinematics mainly contributes to the overall robot inaccuracy in terms of elasticity and backlash [4, 9]. Due to the angular measurement at the drive side, this inaccuracy cannot be compensated sufficiently. If the torque transmitted by a gear changes its algebraic sign the backlash is most distinctive. Even with the use of modern high-precision gears this backlash cannot be avoided completely. A schematic head curve of a gear [11] revealing the relation between torque and torsion angle, is given in Fig. 1. In the zone of high torques the curve is favorable with regard to backlash and stiffness.

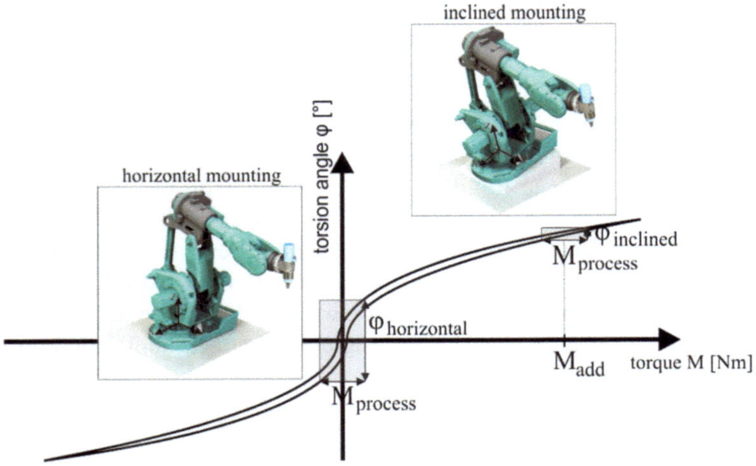

Fig. 1 Schematic gear head curve with operating zones for horizontal and inclined mounting

In order to use one of the favorable zones an artificial torque M_{add} has to be added to the torque resulting from the acting process loads $M_{Process}$. The additional torque always has to exceed the process torque. Because of the requirements of robot based machining concerning precision, the range of loads applied is usually not used to its full capacity, hence the application of the robot is not limited by this concept. The second and third rotational axis of serial robot kinematics are subject to the gravitational force of the attached IR structure, which causes additional torques acting on the respective gears [12]. Since the first axis of horizontally mounted IR is parallel to the direction of the gravitational force no additional torque is applied to the first gear. To achieve this, the robot is set up on an inclined base.

Experimental Setup

The accuracy tests were conducted using an ABB IRB6660 robot. For all experiments a horizontal and an inclined (15° about y-axis) setup were compared. The exemplary analyzed working point in this publication is shown in Fig. 2. Additionally analyzed working points showed similar results. The working plane for all experiments was the x–y-plane.

To examine the performance of both configurations for milling, circles with different radiuses in both rotational directions were machined (Table 1). For this purpose, sheets of CFRP fabric with a thickness of 5 mm and polycrystalline diamond (PCD) end mill cutters with a diameter of 8 mm and two teeth at full engagement were used. The circularity deviation was measured using a Mitutoyo BHN 305 coordinate measuring machine.

To distinguish the occurring effects static stiffness and dynamic circularity tests were conducted. For analyzing the stiffness the displacement of the tool center point (TCP) in x- and y-direction was measured with a Mitutoyo 2046F tactile gauge. Forces, rising and falling in steps of 10 N, between 300 and −300 N were applied to the TCP in the respective direction, by means of a manually adjusted mechanism using a PCE-FG500 force measuring device. With the circularity test the circular path accuracy of TCP movements without external forces was determined. A Renishaw QC-20 W Ballbar caliper was used to acquire the actual radius

Fig. 2 Analyzed working point

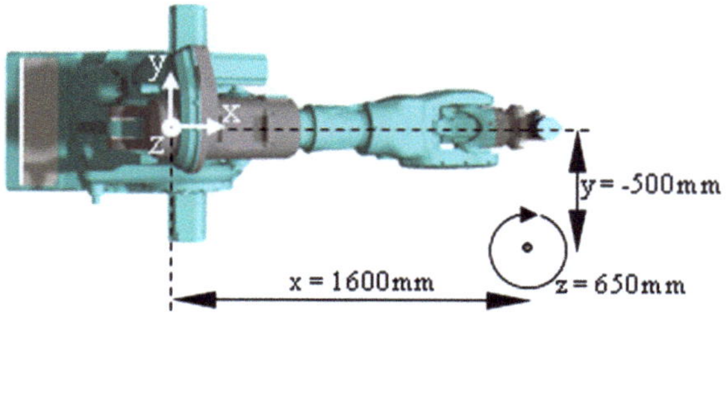

Table 1 Experimental process parameters (with applied forces at stiffness test and average measured process forces at milling test)

	External force		Circular movement		
	x (N)	y (N)	Radius r (mm)	Feeding speed v_f (mm/min)	Spindle speed (1/min)
Stiffness test	−300 to 300		–	–	–
Circularity test	–	–	50; 100; 250	1000; 4000	–
Milling test	150	150	6; 12.5; 25; 50; 100	1000	20000

continuously. Again different radiuses in both rotational directions with two feed rates were analyzed. The exact process parameters for all experiments are shown in Table 1.

Results and Discussion

The results of the stiffness tests are shown for acting forces in y-direction comparing a horizontal and an inclined robot mounting (Fig. 3a). As expected, the horizontal assembly exhibits a distinctive backlash in the range of low forces, i.e. between 30 and −30 N, which disappears at the inclined setting. Furthermore the gradient of the curve is slightly reduced for the inclined mounting. In order to compare the stiffness, the curves were evaluated at minimum and maximum load, showing a stiffness of the inclined setting which is improved by 22 % in negative y-direction and even 28 % in positive y-direction respectively. The described effects derive from shifting the operating point of the first gear to more favorable zones of the gear head curve using the inclination of the robot, as explained above.

The measured displacements in x-direction are smaller than in y-direction for both mountings (Fig. 3b), due to the fact that at this position a lower torque M_Z is

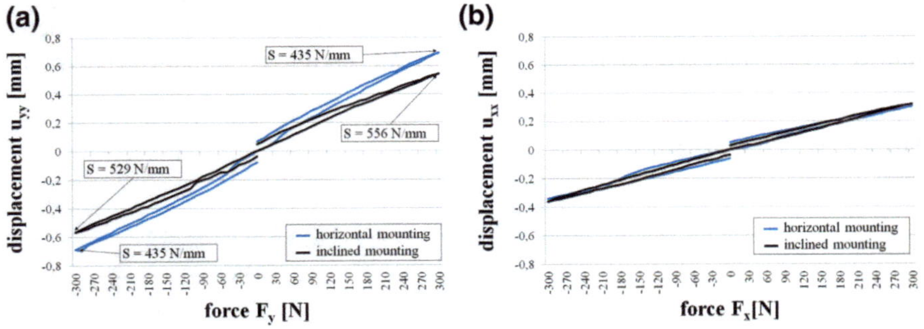

Fig. 3 Displacement curves with force, displacement and stiffness S in **a** y- and, **b** x-direction

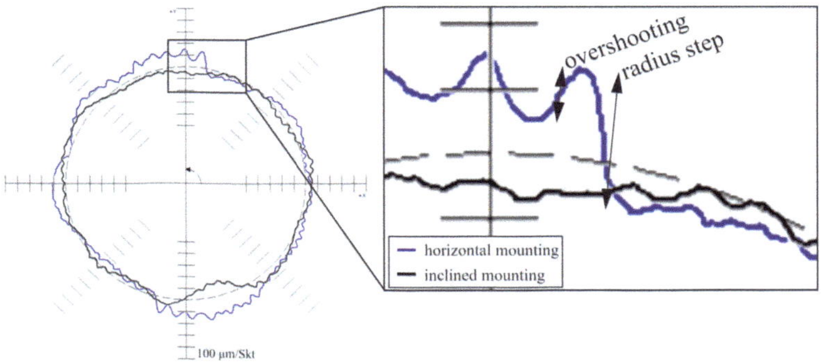

Fig. 4 Circularity with radius steps and overshooting (radius r = 50 mm, feed rate v_f = 4000 mm/min, counterclockwise rotation)

applied on the first axis. The comparison between horizontal and inclined setting shows no significant change.

When analyzing the radius development in the circularity test it can be observed, that the circularity for the horizontal mounting shows distinctive outward radius steps in y-direction, as well as a subsequent overshooting (Fig. 4). This effect occurs exactly at the point where the first axis changes the direction of its movement. Due to the backlash in this particular gear this leads to a delay in the reversal of direction, while axes 2 and 3 continue their movement at maximum speed, which in total causes the observed radial steps. The reduction (elimination) of these radius steps can be ascribed to the shifting of the gear working point of axis 1 and thereby the reduced backlash in this gear. The reduction of overshooting seems to show a similar behavior, especially at high feed rates.

The machined circles show a noticeable reduction in their circular shape deviation (difference between maximum and minimum radius) for all diameters and both circular feed directions (clockwise CW and counterclockwise CCW), when comparing the horizontal and the inclined mounting (see Fig. 5a). The measured process forces oscillate between approximately 150 and −150 N in x- and y-direction, depending on the current feed direction, which leads to constant forces in feed normal direction (65 N for horizontal and 68 N for inclined mounting). For the investigated parameter combinations, improvements in circular shape deviation of upto 0.25 mm could be reached. The average improvement was about 35 % with a maximum of over 50 %.

For all measurements a well-defined oval shape in 45° or 135° direction can be observed (example in Fig. 5b). This derives from the fact that the direction of the feed normal force changes continuously, related to x- and y-coordinates along the circular track. In combination with the above described different IR stiffness characteristics in x- and y-direction this results in the oval shapes. Based on the increase of stiffness with inclined IR mounting, this effect is reduced respectively.

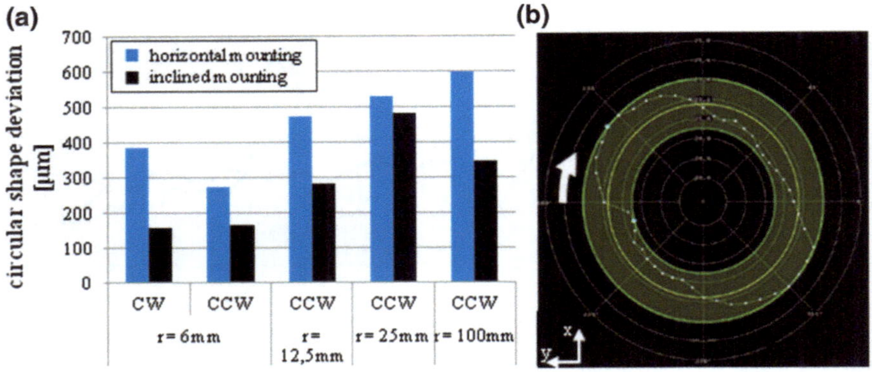

Fig. 5 **a** circular shape deviation of machined *circles*, **b** *oval* shaping for r = 25 mm, horizontal mounting, CW

Conclusions

The effectiveness of a new approach for significantly improving the workpiece quality in milling processes by using an inclined mounting of the IR is proven by the conducted experiments. The following improvements could be achieved:

- displacement reduction of up to 0.15 mm (20 %) at static forces of 300 N in y-direction
- reduction of circular shape deviation (radius steps in y-direction and overshooting) by up to 0.21 mm in dynamic circularity test
- reduction of circular shape deviation of up to 0.25 mm (50 %) in CFRP milling.

It was demonstrated that these improvements can be ascribed to a backlash reduction in the gear of axis one. Current research is aiming for better methods to calibrate the inclined mounted IR, as well as detailed measurements of the behavior of each component in different load situations. Furthermore the effects of different inclination angles, different robots, other workspace areas and the possible durability are under investigation.

References

1. Reinforced Plastics J: Growing demand for eco-efficient aircraft. Reinf. Plast. J **56**(6), 36–37 (2012)
2. DeVlieg, R., Sitton, K., Feikert, E., Inman, J.: ONCE robotic drilling system. In: Proceedings of the 2002 Aerospace Automated Fastening Conference and Exhibition (Chester, England, ISBN: 0768010829), pp. 9–16 (2002)

3. Denkena, B., Immel, J., Schönherr, M.: Industrieroboter für spanende Bearbeitung: Leistungssteigerung hochdynamisch angeregter serieller Strukturen durch Adaptronik. wt Werkstattstechnik online **101**(9), 617–622 (2011)
4. Weigold, M.: Kompensation der Werkzeugabdrängung bei der spanenden bearbeitung mit Industrierobotern. Shaker, Aachen (2008)
5. Wollnack, J.: 3D/6D-Visionsysteme in der Robotik. wt Werkstattstechnik online **97**(9), 718–725 (2007)
6. Abele, E., Bauer, J.: Kamerabasierte Bahnkorrektur für das Fräsen mit In-dustrierobotern. wt Werkstattstechnik online **100**(9), 705–709 (2010)
7. Puzik, A.: Genauigkeitssteigerung bei der spanenden Bearbeitung mit In-dustrierobotern durch Fehlerkompensation mit 3D-Piezo-Ausgleichsaktorik. Jost Jetter, Heimsheim (2011)
8. Zäh, M., Rösch, O.: Steigerung der Arbeitsgenauigkeit bei der Fräsbearbeitung mit Industrierobotern. Zeitschrift für wirtschaftlichen Fabrikbetrieb **106**(9), 658–662 (2011)
9. Abele, E.: Spanende Bearbeitung mit Industrierobotern: Forschungsprojekt ADVOCUT: Entwicklungen und Industrietransfer. Technische Universität Darmstadt, Darmstadt (2007)
10. ABB: Flex finishing: force control for machining applications (2011)
11. Rosenbauer, T.: Getriebe für Industrieroboter: Beurteilungskriterien, Kenndaten, Einsatzhinweise. Shaker, Aachen (1994)
12. Gerstmann, U.: Robotergenauigkeit: Der Getriebeeinfluss auf die Arbeits- und Positioniergenauigkeit. VDI-Verlag, Düsseldorf (1991)

Production of Customized Hybrid Fiber-Reinforced Thermoplastic Composite Components Using Laser-Assisted Tape Placement

C. Brecher, M. Emonts, J. Stimpfl and A. Kermer-Meyer

Abstract The aim of Fraunhofer IPT is to fully exploit the potential of fiber-reinforced thermoplastic composites based on prepreg technology and to achieve at the same time a broad industrial application. Investigations at Fraunhofer IPT have proven that the use of laser-assisted tape placement enables high lay-up rates of thermoplastic prepregs (tapes) with in situ consolidation regarding the local stiffening of plastics and composites. After the optimization of this process for the manufacture of carbon fiber-reinforced components Fraunhofer IPT developed and successfully demonstrated strategies to manufacture advanced composite components via laser-assisted tape placement that combine different polymers and fiber-reinforcements. By the possibility to apply fully automated the most beneficial material in the place, orientation and quantity that's needed, a cross-industrial applicability of this process is ensured.

Introduction

Nowadays, most advanced aircraft structures are made of carbon fiber-reinforced plastics (cfrp) based on prepreg (pre-impregnated fiber) materials and are consolidated as well as cured using autoclaves. Industrial state of the art is the automated processing of unidirectional reinforced prepregs via tape laying of single broad tapes (e.g. prepreg width of 300 mm) for the manufacture of 2D or contoured structures, like wings as well as vertical and horizontal tail planes. For applications where more complex lay-up paths are required, multiple narrow tapes (e.g. 12 or 6 mm in width) are used. Based on the better drapeability and the possibility of individual tape steering this method is suitable for the manufacture

C. Brecher · M. Emonts · J. Stimpfl · A. Kermer-Meyer (✉)
Fraunhofer Institute for Production Technology IPT, Aachen, Germany
e-mail: alexander.kermer-meyer@ipt.fraunhofer.de

B. Denkena (ed.), *New Production Technologies in Aerospace Industry*,
Lecture Notes in Production Engineering, DOI: 10.1007/978-3-319-01964-2_17,

of sophisticated curved structures. Such multi-tape (tow, fiber) placement systems are amongst others used to produce entire fuselages or fuselage segments for commercial and military aircrafts [1–5].

Disadvantages of the current tape laying and fiber placement approaches are the high production costs based on the time and energy intensive curing process after the lamination.

The disadvantages of the state of the art prepreg lamination approach are solved by the automated laser-assisted processing of thermoplastic instead of thermosetting prepregs (tapes). Because of the possibility of an in situ consolidation during the part production without post consolidation and curing steps, all the resources needed for e.g. autoclaving are saved. The shape complexity is additionally enhanced by exploiting flexibility in terms of fusibility (e.g. placement of local-reinforcements, joining with other components), re-shape ability (e.g. thermoforming) as well as further processability (e.g. over-molding). In addition thermoplastic composites enable higher temperature loads (e.g. PEEK, PEKK, PES), excellent chemical properties (e.g. PPS) and in a significant increased damping behavior [6, 7]. The huge variety of thermoplastic matrix systems (in addition e.g. PA, PP, HDPE, PEI) and fiber-reinforcements (e.g. carbon fibers [cf], glass fibers [gf]) enable also other industrial branches (e.g. automotive, oil and gas, wind energy, chemical, civil and mechanical engineering) to realize an optimal solution in respect to the required performance to cost ratio and therefore a cost-effective utilization of thermoplastic composites.

Over the last decade laser systems have been proven to be the most energy-efficient, precise and cost-effective heating method for tape placement in a lot of benchmarks [8–11].

Laser-Assisted Tape Placement

During laser-assisted tape placement the thermoplastic prepregs (tapes) are unreeled from a tape spool and fed by a feed unit into the nip point, see Fig. 1. The surface of the fed tape as well as the one of the already laid down material get heated up over melting temperature shortly before contact with each other using laser radiation. Applying a compaction force using an optionally heated or cooled consolidation roller, the melt up tape gets completely consolidated in situ with the part to be produced without the need for subsequent curing steps. The main difference between laser-assisted tape placement and winding is the use of a rotating mandrel in the case of tape winding, see Fig. 1.

Challenges in the development of a laser-assisted tape placement unit are the development of suited laser optics, temperature control systems and consolidation systems in conjunction with the individual application e.g. the to be processed materials, part geometries, required throughput and flexibility. On the process side the main parameters are the applied temperature, process speed and pressure.

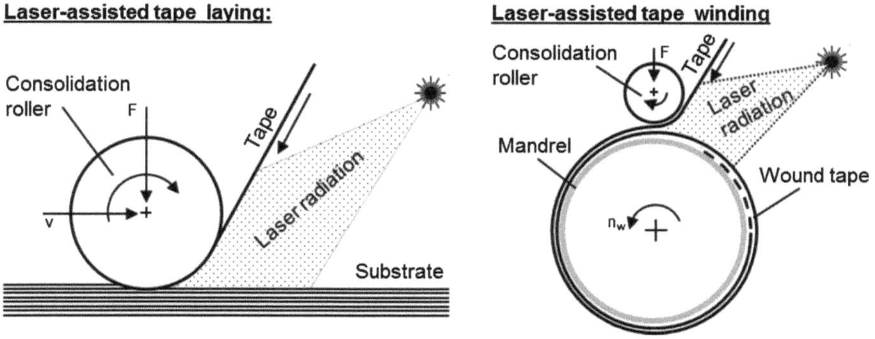

Fig. 1 Principle of laser-assisted tape laying and winding

Evaluations with 3-point-bending tests according to DIN 14125 showed that laminates produced via laser-assisted tape placement have equal mechanical properties as compression molded panels. In experiments with prepregs using a PA 12 matrix system reinforced by AS4 carbon fibers with a fiber volume content of 55 %, it could be proven that 96 % of the bending strength and 104 % of the bending stiffness of the compression molded benchmark laminate can be achieved [12].

Tailored Multi-Material Blanks Made by Laser-Assisted Tape Placement

The exploitation and evaluation of the synergies of laser-assisted tape placement and thermoforming is one goal of the European large scale project » Fibre-Chain « [12, 13]. As highlighted in Fig. 3, composite sheets and parts can be reinforced by laser-assisted tape placement efficiently before and after a forming process to achieve tailored hybrid and multi-material components [14].

It has been demonstrated that in laser-assisted tape placement, composite laminates combined with different types of fibers (e.g. carbon fiber, glass fiber), reinforcements (e.g. fabric, unidirectional [UD]), matrix systems (e.g. PA 6, PA 66, PA 12) and materials with strong varying colors or optical absorption properties can be processed using a diode laser system.

The quality of the welding zone between the base composite laminates and the placed reinforcement tape have been evaluated by a mandrel peel test as shown in Fig. 3. The reinforcements have been produced with a lay-up rate of 300 mm per second.

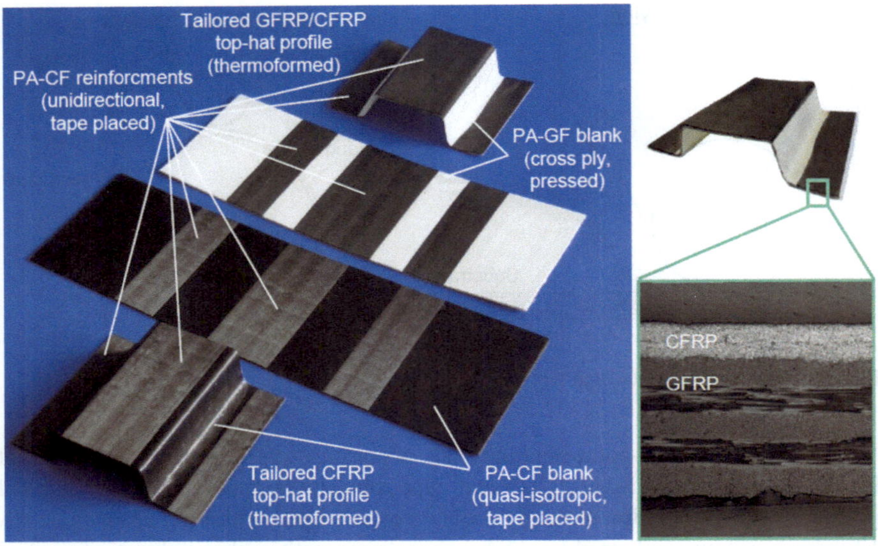

Fig. 2 Tailored parts and tailored blanks via laser-assisted tape placement using a diode laser system

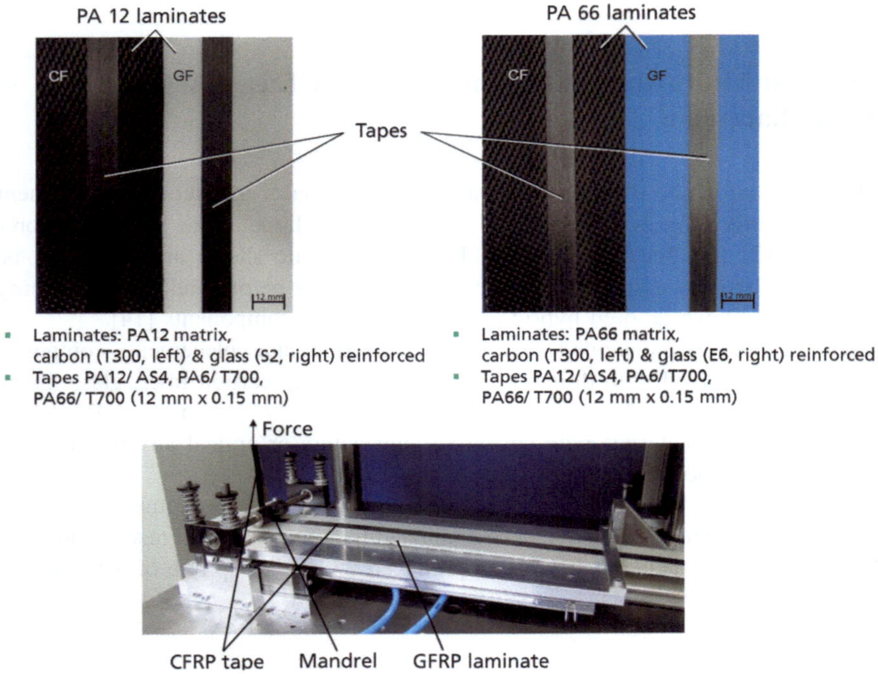

Fig. 3 Overview of evaluation trials

Sheets / Tapes	PA12/carbon	PA 6/ carbon	PA66/carbon
PA 12/carbon	++ (could not be peeled, critical energy always > 8 N/mm)	- (max. 1,7 N/mm)	- (max. 1,6 N/mm)
PA12/glass	+ (max. 4,4 N/mm)	- (max. 1,5 N/mm)	- (max. 1,7 N/mm)
PA 66/carbon	- (max. 2,1 N/mm)	+ (max. 7,3 N/mm)	++ (could not be peeled, critical energy always > 8 N/mm)
PA 66/glass	/ (not tested, low critical energy below 2 N/mm expected)	- (max. 2 N/mm)	- (max. 1, 8 N/mm)

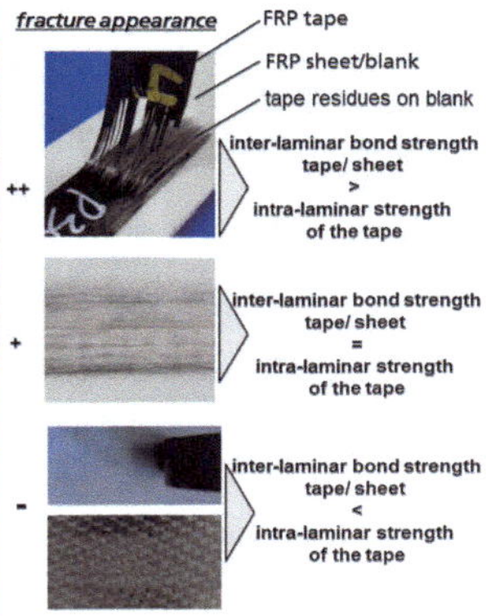

sufficient laser power for tested lay-up rate

insufficient laser power for tested lay-up rate

Fig. 4 Summarized results of reinforcement trials using a diode laser system

Figure 4 summarizes the achieved results of the reinforcement trials using a diode laser-assisted tape placement unit.

The mark "++" indicates a very good weld strength, that exceeded the intra-ply properties of the tape at the tested load case. The mark "+" indicates good adhesion between tape and blank. The inter-ply properties are similar to the intra-ply properties of the used tape at the tested load case. The mark "-" indicates a low joint strength. Furthermore, the critical energy needed to peel the tape is listed.

As expected, the combination of two identical materials did result in the highest weld strength due to similarity of the chemical, thermal and optical properties of the joining partners. When joining different matrix materials it is important to ensure that the chosen process temperature fits to both joining partners. The more thermal stable material needs to be fused without degrading the joining partner that is less thermal stable.

Using an optical heating method the absorption properties of the used material in respect to the used wavelength of the used laser system is important. Different fiber-reinforcements can result in different optical absorption properties. The process can be adapted to joining two materials with different optical properties by adjusting the laser power distribution via the setting of the laser angle.

The reason that the glass fiber-reinforced PA66 (PA6/GF) blank could not be reinforced in the given test plan is because of its low absorption degree in the near infrared spectrum, in conjunction with high welding temperature, high lay-up rate and limited available laser power. Welding a carbon fiber-reinforced PA66 tape (PA66/CF) onto a (PA66/CF) blank at the same process conditions is done with ease, since both joining partners absorb 85–90 % of the used laser radiation close to surface.

A good joint would be achieved by reducing the lay-up rate or by using more laser power. In addition the utilization of carbon black particles within laminate polymer for increased absorption is an option, depending on the application.

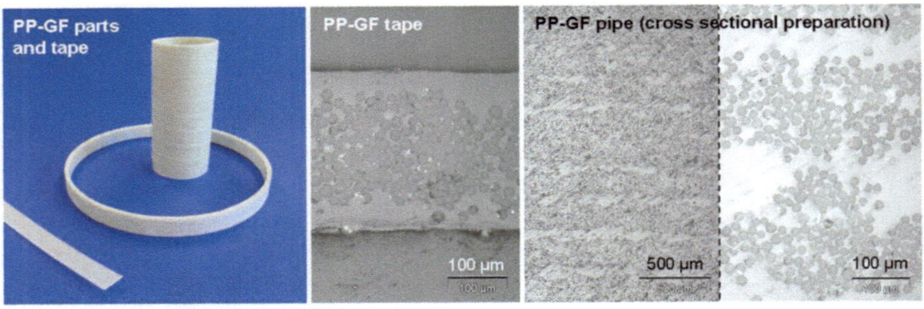

Fig. 5 Parts manufactured by CO_2-laser-assisted tape winding (*left*) as well as cross sectional preparations of raw PP-GF tape (*middle*) and the PP-GF pipe

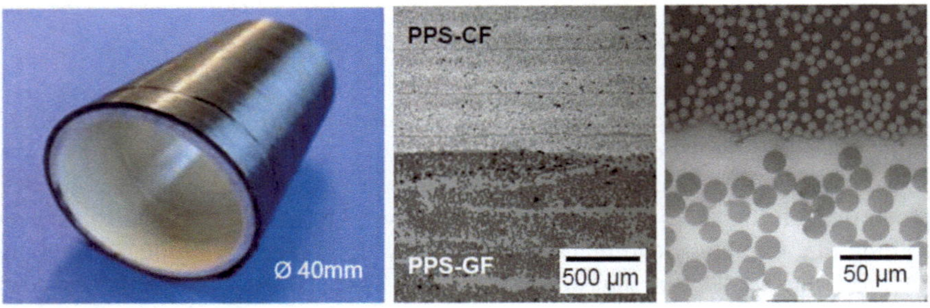

Fig. 6 Hybrid CFRP/GFRP pipe sample produced using CO_2-laser-assisted tape winding

Within the German project » GreenGlassFiber « also the application of a CO_2-laser system for tape winding is investigated, as due to its longer wavelength, the absorption of the laser radiation by the glass fiber-reinforced plastics (GFRP) is high even without carbon black particles/fibers in the material. Figure 5 shows glass fiber reinforced parts with a PP matrix system made by CO_2-laser-assisted tape winding Fig. 6 shows a hybrid part consisting of a gfrp liner surrounded by a cfrp cover with a PPS matrix produced via CO_2-laser-assisted tape winding.

Conclusion

Laser-assisted tape placement with in situ consolidation is a highly productive, automated and flexible method for the lamination of continuous, unidirectional fiber-reinforced thermoplastic prepregs. Since tape placement is a primary shaping process, the production of structural parts, pipes, tanks, tailored blanks and local reinforcement of fabric blanks, composite and plastic parts is feasible. It has been shown that for processing carbon black materials diode laser systems are most efficient and that for milky white materials a CO_2-laser system is beneficial. Hybrid components could be produced with both laser sources with outstanding quality. Additionally it could be demonstrated that multi-material laminates comprising different polymer and fiber-reinforcements in one laminate can be produced efficiently via laser-assisted tape placement with in situ consolidation.

Acknowledgements The authors would like to thank the European Community for funding the EU large scale project "FibreChain—Integrated Process Chain for Automated and Flexible Production of Fibre-Reinforced Plastic Products" at the European Community's Seventh Framework Programme under grant agreement no. 263385. Furthermore, the "GreenGlassFiber" research and development project is funded by the German Federal Ministry of Education and Research (BMBF) within the funding action "SME—Innovative: Resources and Energy Efficiency". The authors are responsible for the contents of this publication. Moreover we want to thank our industrial and academic partners.

References

1. N.N.: Airframers exploit composites in battle for supremacy. Reinf. Plast. **49**(3), 26–32 (2005) ISSN 0034-2617
2. N.N.: Mandrel forms composite fuselage. Reinf. Plast. **49**(3), 24, (2005) ISSN 0034-2617
3. N.N.: One-piece fuselage by fibre placement. Reinf. Plast. **50**(1), 17, (2006) ISSN 0034-2617
4. Edelmann, K., Räckers, B.: Verbundwerkstoffe im A380, Internationaler Kongress, Kunststoffe im Automobilbau, 29.-30.03.2006 Mannheim (2006)
5. Spaeth, A.: Lighter on all sides, in: Aerotec, Ausgabe 5, Dezember 2007, pp. 34–38
6. Gao, S.: Cooling rate influences in carbon/PEEK composites, Part III: impact damage performance. Compos. A Appl. Sci. Manuf. **32**, 763–774 (2001)
7. Bai, J.-M., Leach, D., Cease, S., Pratte, J.: High performance thermoplastic polymers and composites. In: SAMPE, Corina, CA, USA, S. 1391–1405 (2005)
8. Esche, R.V.D.: Herstellung langfaserverstärkter Thermoplastbauteile unter Zuhilfenahme von Hochleistungslasern als Wärmequelle. Diss. RWTH Aachen (2001)
9. Brecher, C., Kermer-Meyer, A., Steyer, M., Dubratz, M., Emonts, M.: Laser-assisted thermoplastic tape laying. JEC Compos. Mag. **47**, 56, S. 71–73 (2010), ISSN 1639-965X
10. Schledjewski, R.: Thermoplastic tape placement by means of diode laser heating. In: SAMPE - Baltimore, MD - 18–21 May 2009
11. Brecher, C., Kermer-Meyer, A., Werner, D., Stimpfl, J., Janssen, H., Emonts, M.: Customized solutions for laser-assisted tape placement, JEC Compos. Mag. **76**, 70–73 (2012)
12. Steyer, M.: Laserunterstütztes Tapelegeverfahren zur Fertigung endlosfaserverstärkter Thermoplastlaminate, Diss. RWTH Aachen (2012)
13. Brecher, C., Kermer-Meyer, A., Dubratz, M., Emonts, M.: Thermoplastische Organobleche für die Großserie, ATZ, Special Karosserie und Bleche, Oktober, S. 28–32 (2010)
14. Brecher, C., Stimpfl, J., Kermer-Meyer, A., Dubratz, M., Emonts, M.: Load-optimised tailored thermoplastic FRP blanks for mass production, JEC Compos. Mag. **64**, 95–97 (2011)
15. Brecher, C., Emonts, M., Kermer-Meyer, A., Janssen, H., Werner, D.: Wirtschaftliche Prozesskette zur funktionsintegrierten Herstellung von belastungsoptimierten thermoplastischen Faserverbund-Bauteilen, Lightweight Design 03/2013, pp. 38–43

Efficient Production of CFRP Lightweight Structures on the Basis of Manufacturing Considerations at an Early Design Stage

B. Denkena, P. Horst, C. Schmidt, M. Behr and J. Krieglsteiner

Introduction

Success in development of lightweight structures is determined by the three disciplines of design, materials, and manufacturing. Focusing on design leads to expensive lightweight structures while overrating production makes it hard to reach structural performance goals. The global optimum of structural performance and cost can only be reached if all three disciplines are equally taken into account. It can be observed that this optimum gets increasingly important for major strategic decisions in lightweight construction industry, e.g. the material concept in future aircraft structures: carbon fiber-reinforced plastics (CFRP) versus aluminum [2, 6].

While development of metallic structures is industrially performed and broadly researched, fiber-reinforced plastics do present new challenges. Design work with homogeneous and isotropic metallic structures is mainly done on a level of part shape and sizing. For composites, the inner heterogeneous and orthotropic structure has to be engineered as well. Therefore, structure development usually has to deal with a higher number of design parameters, raising the need for simulation tools and optimization algorithms. In addition to more sophisticated design procedures, production planning for composite structures gets more challenging as well. For metallic structures, manufacturing usually starts with semi-finished parts having material properties mostly set as in the final product. The material

B. Denkena · C. Schmidt · M. Behr (✉)
IFW—Institute of Production Engineering and Machine Tools, Leibniz Universität Hannover, Garbsen, Germany
e-mail: behr@ifw.uni-hannover.de

B. Denkena
e-mail: denkena@ifw.uni-hannover.de

P. Horst · J. Krieglsteiner
IFL—Institute of Aircraft Design and Lightweight Structures, Technische Universität Braunschweig, Braunschweig, Germany

B. Denkena (ed.), *New Production Technologies in Aerospace Industry*,
Lecture Notes in Production Engineering, DOI: 10.1007/978-3-319-01964-2_18,

properties of composites are mainly determined by the manufacturing processes. Properties such as fiber volume fraction or fiber orientation and imperfections like fiber undulations or inclusions are highly dependent on manufacturing. Therefore process stability has a large impact on structural characteristics. Design mostly accounts for the resulting uncertainties with high knock-down factors for assumed material properties, i.e. decreased lightweight potential [5, 8].

Motivation

Whereas development objectives are commonly measured as a combination of structural performance and cost (e.g. weight savings/production cost), the applied methodologies do not fully serve this goal. State of the art practice is mostly sequential planning, starting with design and then considering processes. Dependencies between both, arising from lay-up, geometry, subdivision of structures, tolerances, and process limits, are not particularly addressed. Interaction is limited to iterations in late design phases when conflicts in production planning get obvious. Since design changes in late phases are disproportionally more expensive than in early phases, the current practice can be considered as inefficient [6].

It is obvious to see that the above mentioned arguments claim a careful attention of design, materials and manufacturing in equal importance right from the start of development. Concerning this, different approaches for multidisciplinary development are available and state of the art. The V-Model for instance defines the cooperation of different disciplines, but focusses only on the product development phase [10]. Simultaneous Engineering suggests increased interaction and parallelization of formerly sequential processes; however, it does not specify detailed procedures [3]. There are more detailed procedures available, but they still do not satisfy the needs of developing composite structures [4, 7].

HP CFK Approach

The new procedure is based on the classical approach described in VDI 2221 [10] (see Fig. 1 and in detail Fig. 2). In a first step, the requirements are defined for the design and the production scenario. After that, the development of a conceptual design follows conventionally.

For the following demonstration, it is assumed that an omega-type section (see Fig. 3) turns out to be most promising in the conceptual design considerations. The phase of embodiment design is split up. First of all, a simplified model is created (see Fig. 3b). This model is dimensioned according to the requirements defined without considering manufacturing aspects. Best available material quality, as it can be found in prepreg tapes without any further imperfections, is used in this stage. The resulting preliminary, so far very rough, design can optionally be split

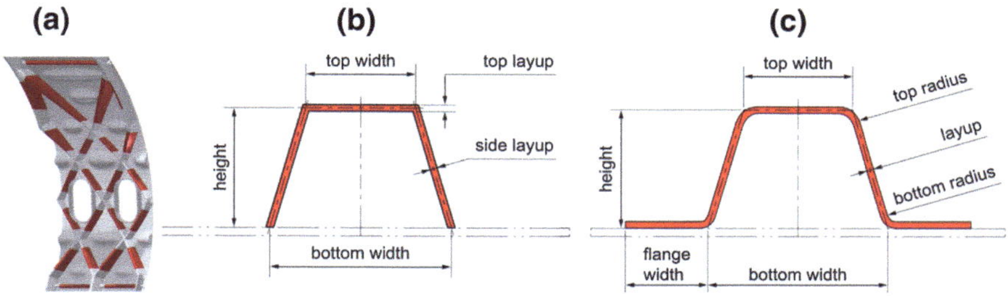

Fig. 1 Development processes: Conventional and HP CFK approach

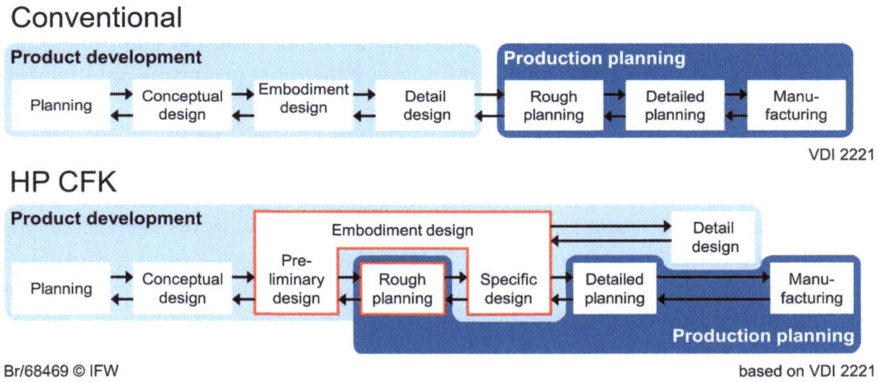

Fig. 2 HP CFK approach in detail

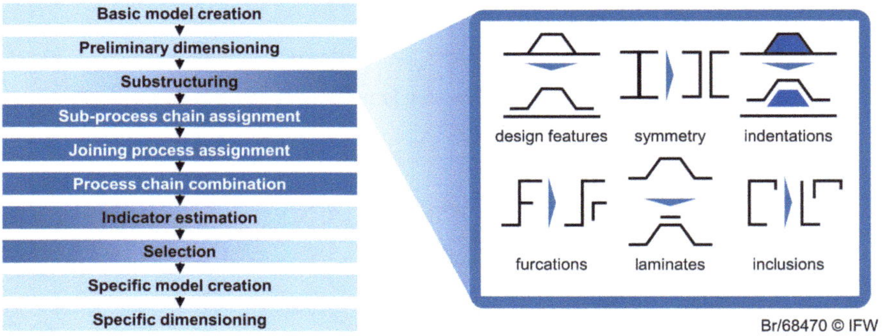

Fig. 3 HP CFK panel and stiffener modeling **a** panel, **b** preliminary design model, **c** possible manufacturing-specific design model

into sets of easier-to-manufacture substructures. The basic criteria for this are defined in a system of substructuring rules using features of the preliminary design such as furcations, hollow sections, laminate properties, or the occurrence of potential common parts. Some are symbolized in Fig. 2 and their application to the sample structure is shown in Fig. 4.

To manufacture each set, in the rough planning phase, different sub-process chains are found and assigned. Each solution contains the four elementary procedural steps to manufacture a fiber-reinforced plastic component in common order of impregnation, forming, consolidation, and curing. In case of sets of multiple substructures, an additional joining process follows. Based on this information, the entire number of sub-process chains and possible joining technologies are combined in a full-factorial pattern.

Processes models were built in advance and production costs of the structures are rated. The weight impact coming with the respective changes in material properties compared to the optimistic preliminary design and the additional mass of joints are estimated. Based on this information the most promising solutions are extracted as visualized in Fig. 5a.

In the next step of embodiment design, models are built based on the process chains found in the last step. The modeling is done in more detail than in the preliminary step (see Fig. 3c). Each model is optimized with respect to the defined requirements and the manufacturing restrictions coming with the respective processes. Finally, the created solutions can be compared by structural performance and cost and the best one is worked out in detail phases. These specific phases are part of ongoing research and will be presented in further contributions. In this regard, a first approach for structural optimization considering interactions of part geometry and draping effects is presented in [1].

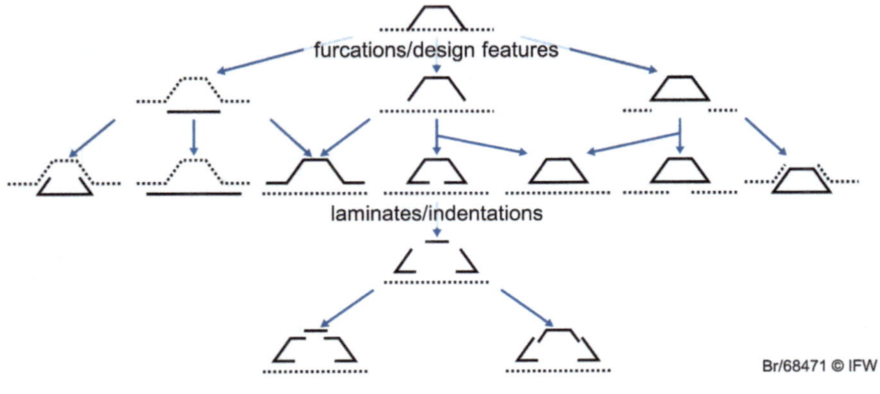

Fig. 4 Substructuring of the preliminary design model

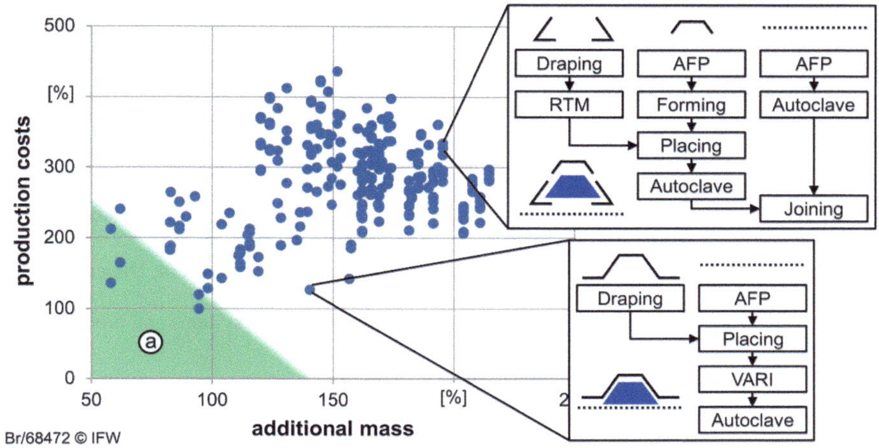

Fig. 5 Estimated cost and weight impact of different solutions and two sample process chains

Conclusion

The presented novel procedure implements a structured way to integrate manufacturing considerations at an early design stage in product development. It generates design solutions, each optimized for a potential process chain, guaranteeing stability and efficiency in manufacturing. The overall solutions of design and production are rated by an evaluation scheme giving the best compromise between cost and structural performance according to given preferences. The material aspect is dealt with in process selection where material types, e.g. fabric, non-crimped-fibers, etc., to be used in specific design are determined.

Bringing design and process chain into agreement, costly, late iterations between production planning and design, well-known from classical development procedures, can be reduced.

Acknowledgments The authors would like to thank the federal state of Lower Saxony and the European Regional Development Fund (ERDF) for financial and organizational support of this project.

References

1. Deniz, O., Horst, P. et al.: Simulation based design optimization of a CFRP fuselage panel according to draping process of carbon fiber textiles using evolutionary algorithms and response surface methods. In: Conference Article, SAMPE, 6–9 May 2013, Long Beach, CA, pp. 1110–1124
2. Eikötter, M.: Synchronisation der Produkt-, Technologie- und Fabrikplanung durch integratives roadmapping. Dissertation, Universität Hannover (2011)

3. Eversheim, W., Schuh, G.: Integrierte Produkt-und Prozessgestaltung. Springer, Berlin (2005)
4. Fallböhmer, M.: Generieren alternativer Technologieketten in frühen Phasen der produktentwicklung. Dissertation, RWTH Aachen (2000)
5. Klammer, J.: Strukturmechanik und Vibroakustik von CFK-Flugzeugrümpfen. Dissertation, TU München (2009)
6. Mazumdar, S.: Composites manufacturing: materials, product, and process engineering. CRC Press, Boca Raton (2001)
7. Müller, S.: Methodik für die entwicklungs und planungsbegleitende Generierung und Bewertung von Produktionsalternative, Dissertation, TU München (2007)
8. Reim, A.: Optimierung von Flugzeugstrukturen mit unsicheren Parametern, Dissertation, TU Braunschweig (2011)
9. Systematic approach to the development and design of technical systems and products, VDI (1993)
10. Design methodology for mechatronic systems, VDI (2004)

Influence of the Fiber Cutting Angle on Work Piece Temperature in Drilling of Unidirectional CFRP

Wolfgang Hintze, Christoph Schütte and Stefan Steinbach

Abstract The increasing use of carbon fiber reinforced plastics (CFRP) for structural components in modern aircrafts also increases the need for their reliable assembly. This is usually accomplished by rivets, which require a large number of precision holes. Drilling is difficult due to contrasting fiber and resin properties and for non-crimp fabrics in particular due to their unidirectional plies. Thermal damage of CFRP components in the vicinity of bore holes should be avoided. Therefore, temperature distributions at the cutting surface of the workpiece as well as in the vicinity of the bore hole have been investigated experimentally while drilling of unidirectional CFRP and confirmed by numerical methods. Temperature changes with fiber cutting angle and heat mainly dissipates in fiber direction.

Keywords CFRP · Drilling · Cutting temperature · Fiber cutting angle · Cutting torque

Introduction

Because of the high specific strength and stiffness, carbon fiber reinforced plastics (CFRP) are important for structural components in aeronautic industry. Assembly of these components is often done by riveting, which requires a large number of holes with high quality in terms of precision and surface integrity. Due to the contrasting thermo-mechanical properties of fibers and resin as well as the orthotropic laminates of CFRP, forces, temperature and tool wear must be controlled in order to avoid decomposition of matrix material or disruption of the surface. These flaws can be very critical in case of non-crimp fabric (NCF)

W. Hintze (✉) · C. Schütte · S. Steinbach
Institute of Production Management and Technology, Hamburg University of Technology, Denickestraße 17, 21071 Hamburg, Germany
e-mail: w.hintze@tuhh.de

B. Denkena (ed.), *New Production Technologies in Aerospace Industry*,
Lecture Notes in Production Engineering, DOI: 10.1007/978-3-319-01964-2_19,

consisting of unidirectional (UD-) laminate layers. Depending on the fiber cutting angle θ between fiber and cutting velocity directions and on the type of drill different modes of damage were found by experiments [1]. Dynamically changing forces and torques have been investigated for UD-laminates during drill rotation [2–4]. The according temperatures are not known up to now.

Highly dynamic measurement of cutting tool temperatures has been done by Ueda et al. [5] using a two-color pyrometer. König et al. [6] used infrared thermography to detect temperatures during CFRP drilling processes. Tool temperature investigations in drilling of fiber reinforced plastics containing woven layers by using thermocouples with an inverted drilling kinematic, i.e. a fixed tool and a rotating specimen were carried out by Weinert et al. [7].

Methods

The temperature distribution at the cutting surface of the component was measured while drilling with a sampling rate of $f_s = 100$ kHz by using a two-color fiber pyrometer Type FIRE-3. The temperatures were detected as the pyrometer optical fiber was put through the coolant channel of the drill. In addition the cutting torque was recorded by a rotating Kistler Type 9123C multi-component dynamometer with a cutoff frequency of $f_s = 1$ kHz, which provides a zero reference mark for synchronization to the cutting temperature. Disc-like UD-CFRP work pieces with an outer diameter of $d_s = 16$ mm were rotating while the drill was kept at rest on the machining table. The setup is shown in Fig. 1.

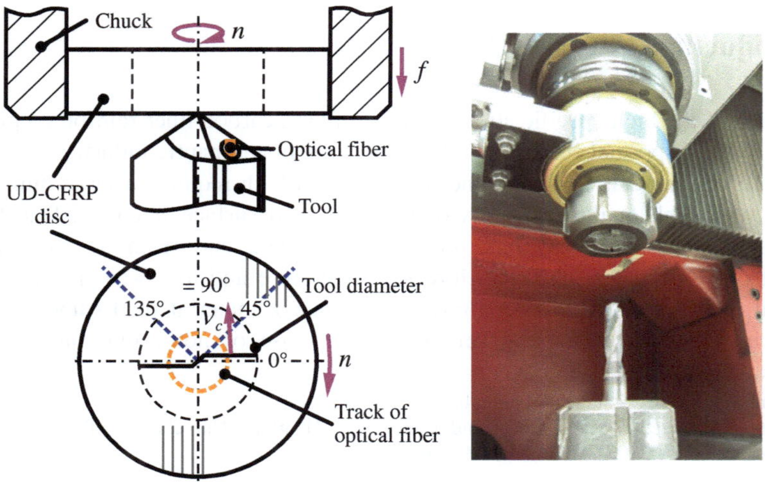

Fig. 1 Experimental setup for measuring cutting surface temperatures

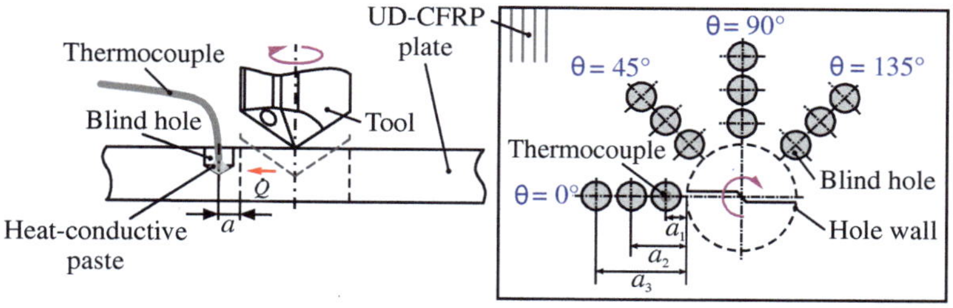

Fig. 2 Experimental setup for measuring work piece temperature distribution

The temperature distribution all around the bore hole was measured using a rotating drill and fixed work pieces with a mantle thermocouple (Type K, diameter $d = 0.75$ mm) in blind holes (depth $t = 2$ mm) at different fiber orientations and distances a to the hole surface. The experimental setup is shown in Fig. 2.

Two different TiAlN-coated carbide drills with a tool diameter of $d = 8.5$ mm, side rake angles of $\gamma_f = 0$ or 25° and a point angle of $\sigma = 140°$ were tested. Cutting edge radii were measured by a confocal microscope Alicona Infinite Focus.

Unidirectional CFRP consisting of CYCOM 977-2 epoxy resin [8] and Tenax IMS60 fibers [9] with a fiber volume ratio of 60 % and a thickness of $h = 4$ mm was used. Three spindle speeds of $n = 2{,}000$, 6,000 and 10,000 rpm and feeds of $f = 0.03$ and 0.05 mm were investigated. The lowest spindle speed leads to an angular resolution of the torque $\Delta\theta = 12°$. The cutting temperature signal was smoothed with the software National Instruments DIAdem 2011 by calculating the central moving average along time intervals $\Delta t = 0.5$ ms.

Work piece temperature distributions all around the bore hole have been numerically calculated using the FEM software Abaqus. The heat generated at the major cutting edge is assumed to distribute in the plane of a single UD laminate ply. Heat exchange between adjacent layers was neglected. A single ply laminate plate of 30 × 30 mm^2 was modeled by its orthotropic properties taken from [8–11] and automatically meshed by 4-node linear heat transfer tetrahedron elements (Type DC3D4) with 36 elements at the hole periphery. The measured cutting surface temperatures during a drill rotation have been represented by an empirical equation and used as boundary condition. Heat dissipation was assumed by surface radiation to the ambient at the peripheral front faces of the plate.

Results of Temperature Measurements and Discussion

First of all, temperature at the cutting surface ϑ_{cs} and torque M_c has been recorded. It is known that the torque varies along the fiber cutting angle θ [2, 3]. Hence, this is also expected for the cutting temperature. The results confirm a big influence of θ,

Fig. 3. The lowest temperature ϑ_{cs_min} is found at a cutting angle of about $\theta = 45°$, the highest ϑ_{cs_max} at an angle of about $\theta = 135°$. Temperature and cutting torque distributions are in phase as illustrated in Fig. 4a.

Different cutting parameters have been tested to investigate the influence of tool geometry, spindle speed and feed on the cutting surface temperature ϑ_{cs} and torque M_c. As shown in Table 1, spindle speed n and feed f have only little influence on cutting surface temperatures ϑ_{cs_min} and ϑ_{cs_max}.

However, the cutting surface temperature ϑ_{cs} is strongly affected by tool geometry and wear. A worn tool with a cutting edge radius of $r_\beta = 35$ µm and a rake angle of $\gamma_f = 0°$ produces much higher temperatures than a sharp one with $r_\beta = 12$ µm and $\gamma_f = 25°$. The radial distance of the coolant channels to the center of the drills was nearly similar, so the influence of different cutting speeds v_c at the analyzed cutting surface can be neglected.

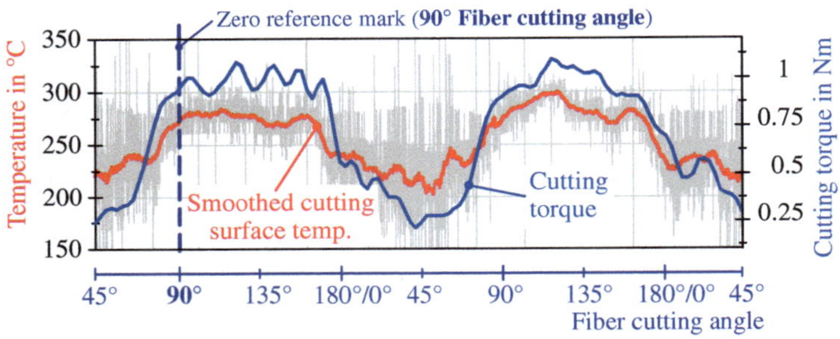

Fig. 3 Cutting surface temperature and torque for drilling UD-CFRP ($n = 2{,}000$ rpm, $f = 0.05$ mm, $d = 8.5$ mm, $\sigma = 140°$, $\gamma_f = 0°$, $r_\beta = 35$ µm)

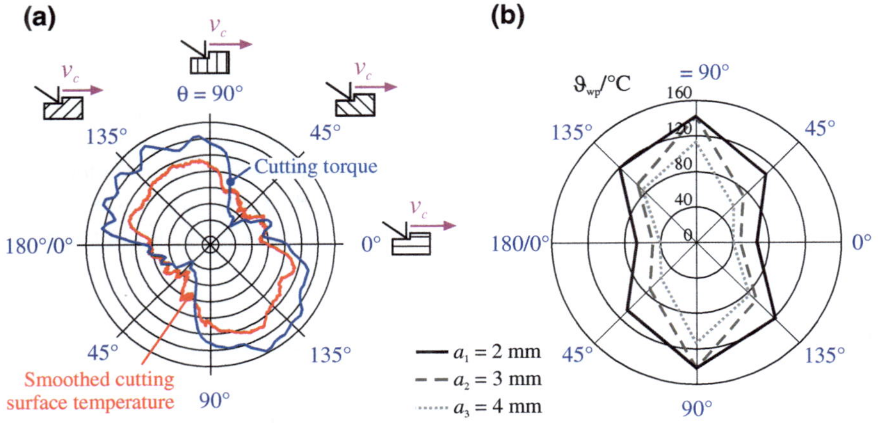

Fig. 4 **a** Cutting surface temperatures and torque, **b** work piece temperature at various distance a for different fiber cutting angles θ ($n = 2{,}000$ rpm, $f = 0.05$ mm, $d = 8.5$ mm, $\sigma = 140°$, $\gamma_f = 0°$, $r_\beta = 35$ µm)

Table 1 Influence of tool, spindle speed and feed on cutting surface temperature and torque ($d = 8.5$ mm, $\sigma = 140°$)

Tool	Spindle speed n rpm	Feed f mm	ϑ_{cs_max} °C	ϑ_{cs_min} °C	M_{c_max} Nm	M_{c_min} Nm
$\gamma_f = 0°$, $r_\beta = 35$ µm	6,000	0.03	370	200	0.6	0.15
$\gamma_f = 0°$, $r_\beta = 35$ µm	6,000	0.05	350	170	0.86	0.29
$\gamma_f = 25°$, $r_\beta = 12$ µm	6,000	0.05	270	100	0.43	0.13
$\gamma_f = 0°$, $r_\beta = 35$ µm	10,000	0.03	350	180	0.6	0.15

Temperature measurements in the vicinity of the bore hole indicate, that the maximum work piece temperature ϑ_{wp_max} occurs at about $\theta = 90°$, the minimum ϑ_{wp_min} at about $\theta = 0/180°$, Fig. 4b. With respect to the heat generation at the cutting surface (Fig. 4a), the heat flow is twisted towards the direction of fibers at distances $a \geq 2$ mm. Because of the thermal conductivities of fibers and resin UD-CFRP exhibits the maximum thermal conductivity in fiber direction and the lowest perpendicular to them [10, 11]. Due to that, the temperature decrease along the distance a at $\theta = 0/180°$, i.e. perpendicular to the fibers is steeper.

Results of Temperature Calculations by FEM and Discussion

The temperature distribution in the vicinity of the bore hole caused by a heat flow from the cutting edge was calculated assuming that the measured cutting surface temperatures are also present at the corner of the drill and therefore, directly at the bore hole surface: $\vartheta_{cs} = \vartheta_{wp}$ ($a = 0$). This assumption is reasonable because of the inferior influence of the spindle speed n on ϑ_{cs}, which accordingly applies for the radial distance of the corner to the center of the drill, see Table 1.

The measured distribution of cutting surface temperatures (see Fig. 3) was approximated as the absolute value of a sine-function with its minimum at $\theta = 45°$ and its maximum at $\theta = 135°$. These temperatures were set as boundary condition of the hole surface temperatures. The simulation time t_{sim} was defined in such a way, that the mean values of measured and simulated temperatures ϑ_{wp} (a, θ) coincide. The calculated time was $t_{sim} = 5.93$ s indicating, that the hole surface was exposed to the temperature not only during cutting but for a longer time due to subsequent friction at the drill periphery. According to a former investigation [3] such friction easily generates a dynamic torque of similar phase and amplitude as the major cutting edges heating up the bore hole surface.

The result is shown in Fig. 5. With increasing distance to the bore hole the isothermals of the temperature field twist towards the fiber orientation which corresponds to $\theta = 90°$. Thus, the FEM analysis confirms the experimental results.

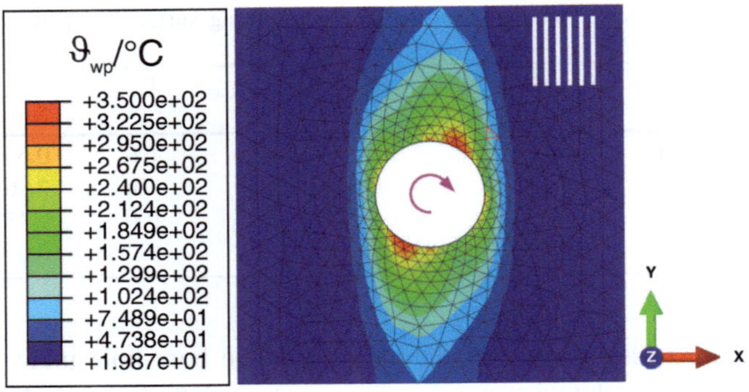

Fig. 5 FE-Analysis (UD-CFRP Layer, $n = 2{,}000$ rpm, $f = 0.05$ mm, $d = 8.5$ mm, $\sigma = 140°$, $\gamma_f = 0°$, $r_\beta = 35$ µm, $t_{sim} = 5.93$ s)

Summary and Outlook

In drilling of UD-CFRP, work piece temperature at the bore hole surface periodically changes with fiber cutting angle. Temperatures mainly depend on tool geometry and wear, reaching values up to 370 °C at the major cutting edge. Generated heat mostly dissipates in direction of fiber orientation in the laminate ply.

The results are useful for industrial drilling of non-crimp CFRP fabrics, because potential thermal damage of bore holes and dynamics of thermo-mechanical tool load can be assessed.

References

1. Hintze, W., Dose, F., Hartmann, D., Schütte, C.: Assessment of surface integrity of machined CFRP—methods and results. In: Proceedings of the 5rd International Conference "Supply on the Wings" (AIRTEC 2010), Frankfurt (Main) (2010)
2. DiPaolo, G., Kapoor, S.G., DeVor, R.E.: An experimental investigation of the crack growth phenomenon for drilling of fiber-reinforced composite materials. J. Eng. Ind. ASME Trans. **118**, 104–110 (1996)
3. Hintze, W., Clausen, R., Hartmann, D. et al.: Precision of machined CFRP: The challenge of dimensional accuracy. In: Proceedings of the 1st International Workshop on Aircraft System Technologies (AST 2007). Shaker Verlag, Aachen, pp. 361–374 (2007)
4. Silberschmidt, et al.: Drilling in carbon/epoxy composites: Experimental investigations and finite element implementation. Compos. A Appl. Sci. Manuf. **47**, 41–51 (2012). doi:10.1016/j.compositesa.2012.11.020
5. Ueda, T., et al.: Temperature on flank face of cutting tool in high speed milling. Ann. CIRP **50**(1), 37–40 (2001)
6. König, W., Graß, P., Willerscheid, H.: Mechanische Bearbeitung faserverstärkter Kunststoffe. Industrieanzeiger **98**, 25–27 (1985)

7. Weinert, K., Kempmann, C.: Cutting temperatures and their effects on the machining behaviour in drilling reinforced plastic composites. Adv. Eng. Mater. **6**, 684–689 (2004). doi:10.1002/adem.200400025
8. Cytec (2012) CYCOM 977-2 Epoxy resin system—technical data sheet. http://www.cemselectorguide.com/pdf/CYCOM_977-2_031912.pdf. Accessed 28 Jun 2013
9. Toho Tenax (2008) Produktprogramm und Eigenschaften für Tenax IMS Filamentgarn. http://www.tohotenax-eu.com/fileadmin/tohotenax/downloads/Produkte/Technische%20Datenblaetter/IMS60_dt.pdf Accessed 28 Jun 2013
10. Schürmann, H.: Konstruieren mit Faser-Kunststoff-Verbunden. Springer-Verlag, Heidelberg (2007)
11. Sheik-Ahmad, J.Y.: Machining of Polymer Composites. Springer-Verlag, Heidelberg (2009)

Increase of Process Stability with Innovative Spindle Drives

W. Bickel, K. M. Litwinski and B. Denkena

Abstract Especially in High Speed Cutting (HSC), the potential of machine tools often cannot be exploited as vibrations at the main spindle limit the productivity. At the Institute of Production Engineering and Machine Tools (IFW) and the Institute for Drive Systems and Power Electronics (IAL) of the Leibniz Universität Hannover, a new active method to increase the process stability is being researched. With the help of an electromagnetic actuator, which was integrated into the drive of a milling spindle, highly dynamic damping forces can be applied to the spindle shaft. Especially with stiff tool/tool holder combinations, the process productivity can be increased significantly hereby. This article describes the implemented concept and shows first promising experimental results of the proposed method.

Introduction

On the one hand, the developments and improvements in cutting material and technology lead to steady rising process parameters in machining. On the other hand, the potential, which is made available by the progress, can often not be exploited as vibration effects at the main spindle tend to destabilize the process and therefore limit the productivity. The literature shows numerous different approaches to this phenomenon. A conventional method to avoid instable cutting processes (e.g. chatter) is the reduction of the machining parameters (spindle speed or depth of cut).

In some cases, different strategies for the variation of the spindle speed showed positive effects on the process stability [1]. Furthermore, approaches to increase

W. Bickel (✉) · K. M. Litwinski · B. Denkena
Institute of Production Engineering and Machine Tools (IFW), Leibniz Universität Hannover, Hannover, Germany
e-mail: bickel@ifw.uni-hannover.de

B. Denkena (ed.), *New Production Technologies in Aerospace Industry*,
Lecture Notes in Production Engineering, DOI: 10.1007/978-3-319-01964-2_20,

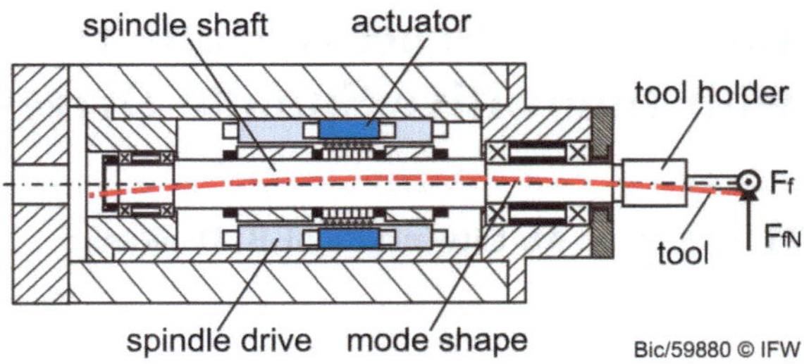

Fig. 1 Principle of the integrated spindle damping

the systems damping are on the rise. Passive methods with a hydrostatic damper at the main spindle [2] or adjusted mass dampers [3] showed good results in stabilizing drilling and turning operations. Active approaches were pursued in [4–6]. Here, actuators were positioned at the main spindle or the workpiece to influence the process dynamics towards stable cutting conditions. The use of a magnetic bearing, positioned next to the conventional ball bearing on the tool side, in [7] also could increase the process stability through damping.

In the project "Method of drive integrated damping of spindle vibrations" a new method to increase the process stability is proposed. It is based on the integration of an electromagnetic actuator into the drive part of a milling spindle. For this purpose, an existing motor spindle concept has been modified to minimize the power losses due to the required space for the actuator. The permanent magnets of the spindle shaft have been replaced by a laminated core in the area of the actuator. Eddy current sensors have been aligned next to the actuator to measure the spindle vibrations. In machining, the actuator can apply highly dynamic damping forces (f_{max} = 1500 Hz) to the spindle shaft and hence reduce undesired vibration effects. An illustration of the concept is shown in Fig. 1. The placement of the actuator between the conventional bearings offers advantages concerning the system dynamics as no external parts have to be mounted to the spindle. Moreover, the overhanging part of the shaft is left unchanged so that the compliance at the tool center point (TCP) is also not reduced in comparison to the original spindle.

Design and Construction of the Damped Motor Spindle

In order to design the method, a finite-element model of a milling spindle has been set up. For this purpose, the dynamics of several spindle components have been measured and unknown parameters of the corresponding FE models were identified on the base of the measurements. The final FE model has been assembled from the component models and further unknown parameters such as damping were estimated. The resulting spindle model showed good correlation with the measurements. This is very important as the efficiency of the method strongly depends on the positioning of the actuator for which detailed knowledge of the system dynamics is necessary. The designed model and a comparison of the measured and

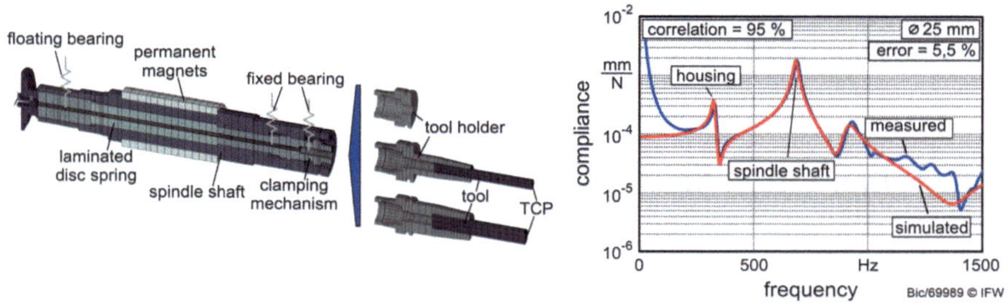

Fig. 2 Finite-element Model of the spindle and comparison of FRFs

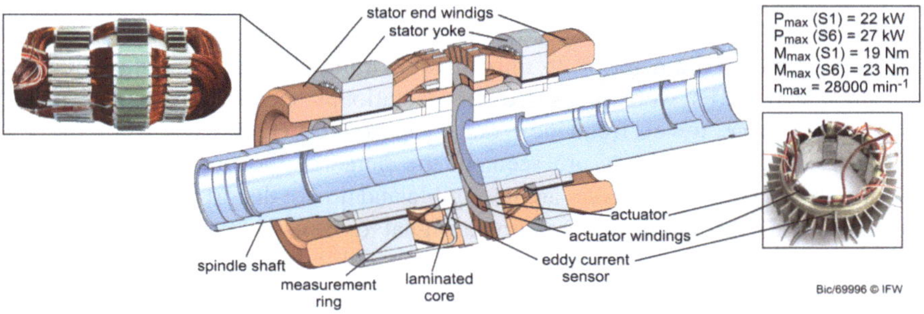

Fig. 3 Design and construction of the spindle prototype

simulated Frequency Response Functions (FRF) at the TCP for the spindle with a tool of 25 mm diameter and a protruding length of 89 mm are shown in Fig. 2.

Based on dynamic simulations where the damping effect of the actuator at different positions has been considered, an actuator system was developed which can be integrated into the spindle drive at the adequate position. Figure 3 displays the design of the novel spindle drive with some technical specifications.

For the integration, the active component of the spindle has been axially divided into two parts. The actuator was placed into the resulting gap. To prevent additional end windings which would result from a separation, the persistent stator windings are guided through the outer diameter of the actuator within a duct system. On the inner diameter, the actuator contains six teeth where two subsequent teeth are connected by one winding. Hence, three separately controlled sub actuators are aligned in sectors of 120° around the circumference. The displacement sensors have been placed in a plastic ring bound to the actuator whereas every sub actuator contains one sensor. The entire ring is positioned beneath the bending radius of the stator windings so that only slight additional loss in space resulted from the sensors. Opposed, a measurement ring made of aluminum has been placed on the shaft to prevent electrical saturation effects and therefore enable precise vibration measurements even in high magnetic fields. For the assembly of the drive, the stator yoke has been as well designed radially divisible. This enables the winding of the stator from the outer radius before assembly of the yoke.

Integration into a Machine Tool

For machining tests, the spindle was integrated into a Heller machining center, whereas the control of the spindle was independent from the machine control. The according measurement setup is shown in the left part of Fig. 4.

For the measurement of the FRFs in first place, eddy current sensors have been aligned in two perpendicular directions at the tip of the engaged tungsten carbide tool dummy. The system was excited with an impulse hammer at the measurement position at different rotational speeds. In the right part of Fig. 4, the recorded FRFs for the tool with a diameter of 25 mm are shown in two orthogonal directions. As the presented FRFs were measured at a rotational speed of $n = 3000\ \text{min}^{-1}$, especially in the lower frequency range up to 400 Hz effects from the tool roundness and concentricity are seen. In the range of the first spindle eigenfrequency (>500 Hz), these effects are negligible and do not distort the measurements. In addition to the undamped reference signal (blue), the measured signals for different damping parameters are plotted in gray color. In the y direction a decrease of the amplitude in the spindle eigenfrequency is observed with increasing damping. Exceeding the red plotted minimum leads to instability in the actuator controller so that no higher reduction of the amplitude was possible. In total, a reduction of 55 % in the amplitude of the resonance was achieved with the damping. Regarding the orthogonal direction, several resonances were found in the corresponding frequency range. Nevertheless, the amplitude in the dominating compliant mode could be reduced by 35 % in the optimum (red). Here effects were found which were also observed in the prior simulations. With exceeding of the optimal damping, the dominant resonance shifts towards a higher frequency and the amplitude rises again (plotted in green). Here the optimal damping constant was found to be 35 Ns/mm which is very similar to the simulated results. Additionally, an eigenfrequency at approximately 350 Hz is visible in the measurements in both

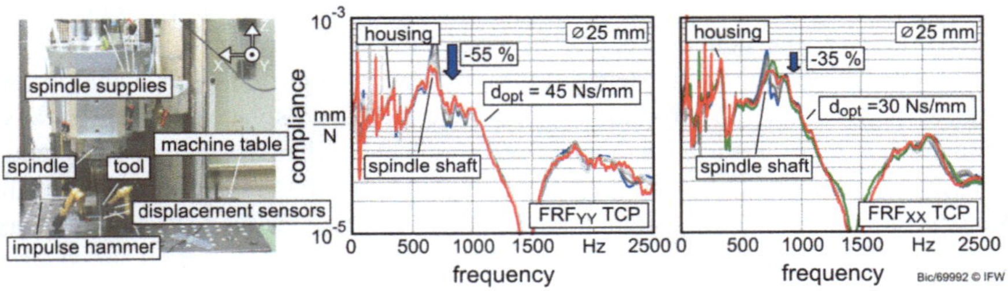

Fig. 4 Measurement setup in the machine tool and measured undamped/damped FRFs

directions. This resonance contains a mode shape, where the entire spindle vibrates in the spindle base so that only a slight damping effect of the actuator is possible.

Experimental Validation of the Damping Characteristics in Machining

The experimental validation of the damped spindle was carried out in extensive milling tests with the aluminum alloy AA7075 and different tool geometries. First of all, the developed damping strategy was evaluated in stable milling conditions. Here, velocity dependent damping forces were applied especially to the separately excited vibrations from the tooth engagements. No significant influence either on the process signals or on the machined surface was found. Considering the process stability, the efficiency of the method was estimated at different instable cutting conditions. Figure 5 exemplarily displays a recorded acceleration signal at the spindle base in time domain as well as the measured displacement signal of the spindle shaft at the actuator in time–frequency domain in x direction for unstable cutting conditions in slotting. Furthermore, the respective machined flank surface is shown.

In the process, the feed was disrupted after approximately 4.5 s and the damping was activated for the rest of the time. Especially in the frequency domain, the signal of the undamped phase is dominated by an eigenfrequency of the system in the x direction at approximately 900 Hz. Furthermore a resonance from the y direction can be found at 700 Hz. With the activation of the damping, a significant decrease of the amplitude in the domination eigenfrequency as well as in the other occurring resonances was achieved, whereas the amplitude in the tooth engagement frequency did not change. As a result, the cutting process became stable.

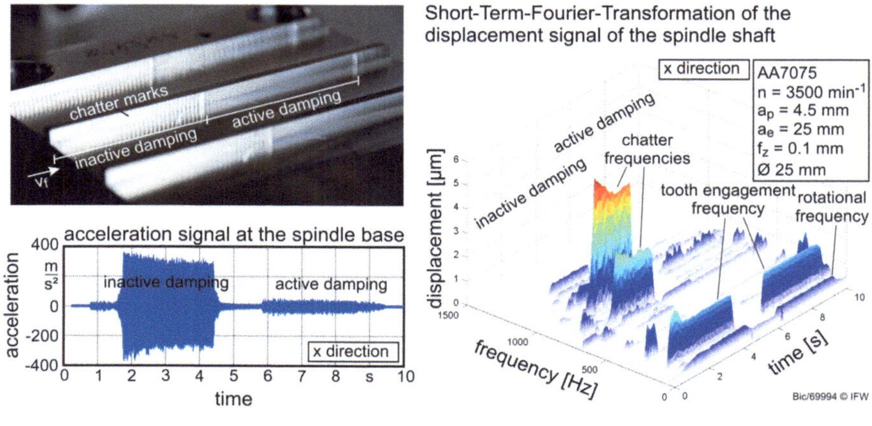

Fig. 5 Stabilizing of an instable slotting process

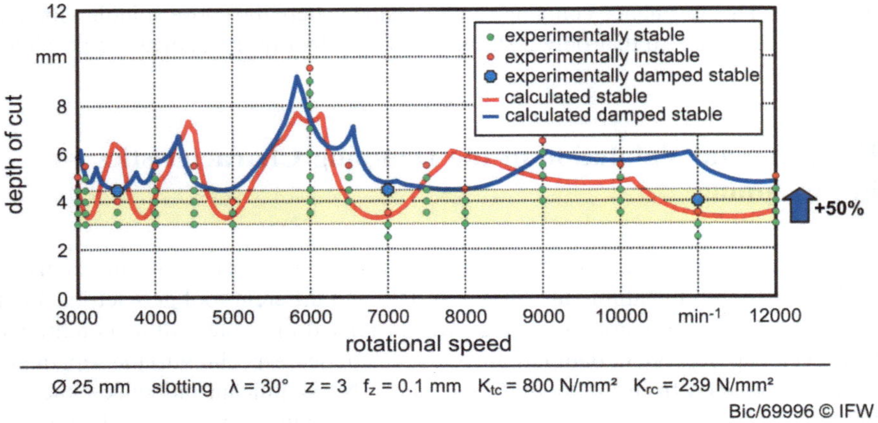

Fig. 6 Experimental and calculated stability lobes for undamped and damped processes

Similar effects were found in the machined flanks. While the instable process left chatter marks, the surface in the stabilized process was only created by the regular tooth engagements and improved clearly. To summarize, stability lobes have been recorded for the described process and are plotted in Fig. 6.

In addition to the experiments with the undamped system (green/red circles), blue circles present the experimentally stable damped cutting limits. Furthermore, solid lines show the calculated stability lobes on the base of FRF measurements in undamped and damped states (red/blue). Although there are differences between the calculated and measured cutting limits, the tendencies can clearly be seen. In specific areas, the stable cutting depth could be increased by 50 % with the introduced damping method.

Summary and Outlook

This paper presents the construction and design of an innovate spindle to increase process stability in milling. An electromagnetic actuator was integrated into the spindle drive to apply highly dynamic radial damping forces to the spindle shaft. First experiments showed a significant decrease in the amplitude of the chatter frequencies in instable cutting conditions due to the damping method. Consequently the stable cutting depth could be increased by up to 50 %.

However, the dynamic system changed its characteristics with the use of different tools or due to the cutting process in the milling tests. The change of the dominating spindle resonance in particular, demands steady adjustments of the damping control in order to keep the system stable. So far, the necessary steps (setting the damping constants or the filter design) were performed manually and were therefore not optimized and prone to errors. In future work, these operations will be automated by the spindle controller. This will also enable the autonomous in-process adjustment of the spindle to the situation. Moreover, the constructed prototype was only a demonstrator for the efficiency of the method. On the way to a fully integrated damped spindle without external parts in the drive, ongoing research will focus on the development of the corresponding new drive part.

References

1. Al-Regib, E., Ni, J., Lee, S.-H.: Programming spindle speed variation for machine tool chatter suppression. Int. J. Mach. Tools Manuf **43**, 1229–1240 (2003)
2. Weinert, K., Hennes, N., Krell, M.: Spindle and toolsystems with high damping. CIRP Ann. Manuf. Technol **48**(1), S. 297–302 (1999)
3. Yang, Y., Muñoa, J., Altintas, Y.: Optimization of multiple tuned mass dampers to suppress machine tool chatter. Int. J. Mach. Tools Manuf **50**, 834–842 (2010)
4. Rashid, A., Nicolescu, C.N.: Active vibration control in palletised workholding system for milling. Int. J. Mach. Tools Manuf **46**, 1626–1636 (2006)
5. Ast, A., Braun, S., Eberhard, P., Heisel, U.: Adaptronic vibration damping for machine tools. Ann. CIRP **56**(1), S. 379–382 (2007)
6. Denkena, B., Will, C.: Schwingungsregelung einer adaptronischen Frässpindel mit Piezoaktoren. wt Werkstattstechnik online 11/12 (2007)

Towards a Cax-Framework for Adaptive Programming Using Generic Process Blocks for Manufacturing

Gunter Spöcker, Thomas Bobek, Lothar Glasmacher and Fritz Klocke

Abstract Over the last years the aerospace industry demands for increasingly complex process chains, as worn, high-value parts have to be repaired and resources have to be used more efficiently. To cope with this growing complexity, tool support ist required for different reasons. The process chains have to be layed out, the knowledge of the processes has to be captured and should be easily available. Furthermore, process chains evolve over time and it should be possible to use the research results to optimize in-use process chains. In this paper it is shown how the function block approach can be extended by "Generic Process Blocks for Manufacturing" (GPBM) in order to meet these requirements for offline adaptive process chains. For this, GPBM introduces a new placeholder function block which supports the reuse and exchange of already modeled process chains. In addition, a case study discusses how this approach can help to transfer a process chain originally designed for tip repair of a single gas-turbine airfoil to a blade integrated disk (BLISK) and how the corresponding process chain knowledge can be stored in a common place.

G. Spöcker (✉) · T. Bobek · L. Glasmacher · F. Klocke
Department CAx-Technologies, Fraunhofer Institute for Production Technology IPT, Steinbachstrasse 17, 52074 Aachen, Germany
e-mail: gunter.spoecker@ipt.fraunhofer.de

T. Bobek
e-mail: thomas.bobek@ipt.fraunhofer.de

L. Glasmacher
e-mail: lothar.glasmacher@ipt.fraunhofer.de

F. Klocke
e-mail: fritz.klocke@ipt.fraunhofer.de

B. Denkena (ed.), *New Production Technologies in Aerospace Industry*, Lecture Notes in Production Engineering, DOI: 10.1007/978-3-319-01964-2_21,

Introduction

In the last years process chains in aerospace industry have become increasingly complex and are expected to achieve higher quality at the same time. Reasons for this are the demand to repair high-value parts (e.g. blade integrated disks—BLISKs), the request for resource efficiency and new approaches such as individual manufacturing—in Germany known as "Industry 4.0". Modeling and planning these process chains requires detailed process knowledge and experienced personnel, since this is still mostly done manually. Current CAD/CAM applications used in industry are operating on the nominal geometry and mainly using tree-based structures that describe the manufacturing sequence of the workpiece. This hinders the design of adaptive processes with varying process flows since dependencies between process steps are not directly apparent in this kind of structure. Furthermore, knowledge is not centrally gathered, as the process chains are usually stored in the part or assembly file.

This paper introduces "Generic Process Blocks for Manufacturing" (GPBM) by extending a function block approach with parameterized types in order to support programming offline adaptive process chains, capturing the relevant knowledge and making it easily available for future programming jobs.

This approach allows defining subgroups of manufacturing process chains with placeholders, which can be instantiated and used as modules to form specific process-chain blocks. Therefore, knowledge can even be captured across various process technologies in these partially-defined process templates and leveraged for new products as it will be shown in the case study of this paper. This approach was studied and successfully tested based on a software platform called "CAx-Framework" developed by the CAx department of Fraunhofer IPT within the Fraunhofer Cluster of Innovation "TurPro" (Integrative Production Technology for Energy-efficient Turbomachinery) [1].

State of the Art

Modeling and planning of process chains has already been addressed in research work in the context of computer-aided process planning (CAPP). CAPP tackles the problem of finding and selecting an ordered combination of available manufacturing processes to enable the successful manufacturing of the desired product. Different approaches have been studied to design complex process chains such as object-orientation, petri-nets, feature recognition and knowledge-based approaches [2]. In any of these approaches a modeling language was used to express the process chain and two major approaches have been proposed: STEP-NC and function blocks.

STEP-NC is a machine tool control language based on the STEP standard for the exchange of products (ISO 10303) and contains all information for manufacturing, through the description of machining entities, workingsteps, workplan, tools, machining strategies, etc. [3]. Each of this data is described by a corresponding application protocol which itself is defined in the data modeling language EXPRESS. STEP-NC describes data structures and does not include any intelligent functionality itself [4]. Typically, STEP-NC forwards meta information through the whole process chain. The idea behind this approach is that the CNC controller at the machine tool is expected to know best how to produce a specific feature, since it can leverage its detailed information of the kinematics. The future goal of STEP-NC is to support online adaptive CNC controllers, which can optimize machining parameters and tool-paths in real time [5]. However, STEP-NC is limited to feature-based approaches as it is only descriptive.

The second commonly proposed approach to program process chains is by leveraging function blocks (IEC 61499) [6]. Function blocks were originally designed for distributed industrial-process measurement and control systems but have also been applied to process planning, scheduling and execution [7]. Each function block defines sets of events, data in- and outputs and an execution control chart (ECC—an extended form of a state machine). On receiving an event, the ECC potentially changes its current state, may start the execution of a specific algorithm and is able to emit an output event, which in turn can trigger the ECC of another function block. In contrast to STEP-NC this approach is not limited to feature-based manufacturing since a function block can contain any type of logic.

Recently, a combination of the two previously described approaches was proposed, using function blocks to implement a STEP-NC processor on the CNC [8] and function blocks with STEP-NC as data model to facilitate data exchange among heterogeneous systems [9].

Function Block Approach

As this paper aims towards offline programming of adaptive process chains and the preservation of the involved knowledge, the function block approach is used as a basis. This decision was made due to the features the function block approach provides: (i) Algorithms, applications or resources can be integrated abstractly in the function block design, helping to interconnect the various systems used in the heterogenous environment of adaptive process chains. For example this option might be required to extract deviations of measurement results from nominal CAD geometries with a specialized software and feed them to the next process step automatically. (ii) It enables an easy integration of user interactions in an event-driven architecture, e.g. for machine tool changes or verification steps. (iii) It is also appropriate for industries which do not want to reveal their intellectual property since binary distribution is possible [10].

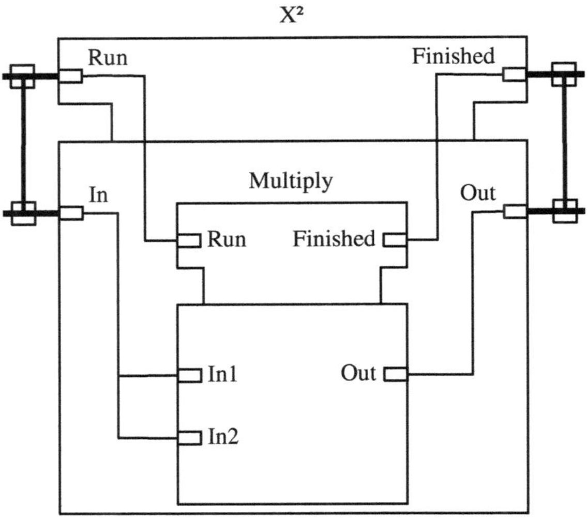

Fig. 1 Composite function block implementing x^2 by using a basic function block for multiplying two input values

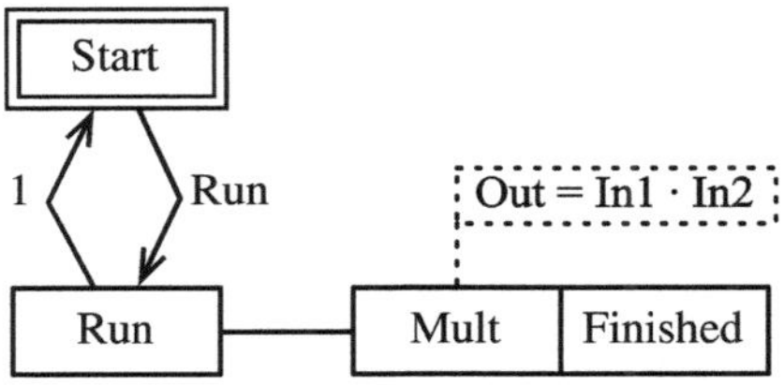

Fig. 2 Example ECC for calling the multiplying algorithm

IEC 61499 defines three types of function blocks: The basic function block can contain internal variables and algorithms, which can be triggered by the ECC to operate on given input data. The second type—composite function block—consists of other function blocks. The third—service interface function block—is ignored in this paper as it used for direct hardware communication.

Figure 1 gives an example of a composite function block for calculating x^2. On receiving the event "Run" the data of the input port "In" is read from the input buffer (visualized by the vertical line and the connection squares, also called WITH construct) [10] and forwarded to the ports "In1" and "In2" of its child block. This basic function block implements the ECC shown in Fig. 2. The event forces the ECC to transit to state "Run", trigger the multiplying algorithm, send the output event "Finished" afterwards and then return to the original state.

However, there is a drawback in the function block approach: It can only describe completely defined and fixed composite function blocks. Together with the definition of a composite function block each of its children has to be defined, all control paths of the event logic and data interfaces have to be fully known as well. This characteristic is not a limitation in the standard use cases of IEC 61499. In adaptive CAx process chains the possibility to exchange parts of a process chain is very useful for the centralization of knowledge because many process chains are partially similar to each other in terms of the working steps involved. For example, most manufacturing process chains start to determine the position of the raw part in a machine tool but the steps after that will vary depending on the specific process plan. This approach requires manual copying of constant function blocks by the operator and manual adaption of varying parts of the process chain. Subsequently, changes in the different copied versions will result in an unmanageable database of function blocks. The same argument applies to avoiding duplicates of code bases [11].

Alternatively, the composite block can include all options and allows a selection based on input data or the state of internal variables [12]. A growing number of options makes this alternative less applicable, especially if there are chains of such alternative sets. Another problem with this approach can be noticed if the alternatives are constantly growing (e.g. if a new manufacturing process gets available).

This problem will become especially eminent in complex composite function blocks as they occur in adaptive process chains. Furthermore, those will be constantly enhanced and the corresponding evolving knowledge is important to gather. Information about the revision history of a process is important later on and is much easier visible if changes can be reduced to the minimum in one function block.

The GPBM

To overcome these problems, the GPBM approach proposed in this paper introduces a new type of function block "placeholder", which can be used in composite function blocks. The basic idea is to provide exchangeable function blocks with compatible interfaces. It requires already defined event interfaces and data in- and outputs. An example of such placeholders is shown as gray function blocks in the example process chain in Fig. 3. The details of this process chain will be discussed in the case study. So far it is important to notice that the placeholder function block can be used like any other function block type but without the requirement to define what is happening internally. Although, specifying an execution control chart and sub function blocks is possible and will be replaced by the actually placed function block. Such a "default implementation" might simplify debugging and developing of the composite function block itself. This concept is very similar to parameterized types and generics [13]. As the main advantage process-related changes in the GPBM will automatically update all dependent processes without further user interaction.

Placeholder function blocks can be instantiated by specifying a function block which should be put in the position of the placeholder function block. This instantiation is illustrated in Fig. 4 by the gray shapes in the form of a function

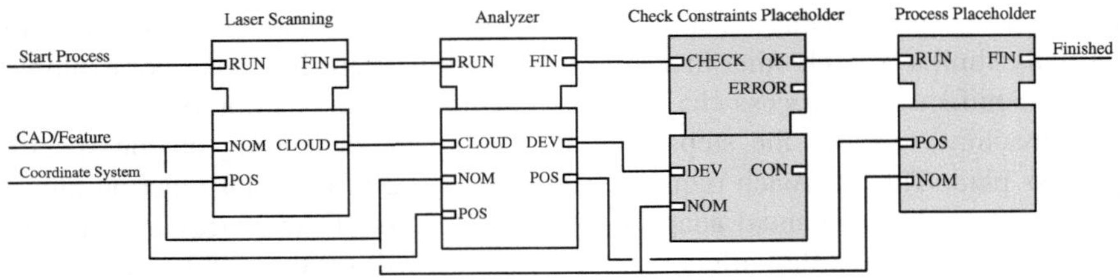

Fig. 3 Generic blade process step using the newly introduced GPBM

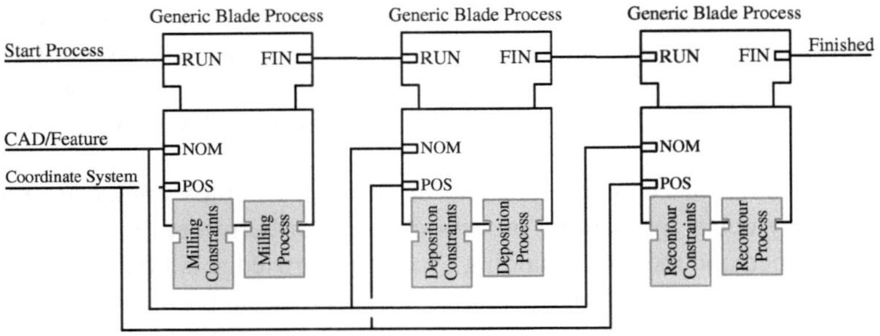

Fig. 4 Blade tip repair process chain using instantiated generic function blocks

block. In order to allow a function block to be placed in a placeholder, it has to meet two conditions:

1. The event and data inputs of the placeholder function block must be a superset of those of the placed one (including the WITH constructs)
2. The event and data outputs of the placed function block must be a superset of those of the placeholder one (including the WITH constructs).

If one of the two conditions is not met, the underlying tool chain shall report an error, as behavior would be undefined. A single composite function block is allowed to contain multiple placeholder function blocks. Furthermore, there is no limitation in respect to the type of function block to be placed in such a placeholder function block. In particular even a placeholder function block of the parent composite function block is allowed, resulting in nested generics.

This approach is quasi-compatible with the existing IEC 61499 standard since an instantiated placeholder function block can be converted to one of the standard function block types. This can be done with a preprocessor either integrated into the function block framework/tool chain or implemented as a standalone software tool. The proposed GPBM are compatible with distributed systems, as the placeholder can be expanded to the corresponding fully defined function block net.

A Case Study

To demonstrate the benefit of the proposed extension compared to standard IEC 61499, a simplified adaptive process chain for tip repair of a BLISK will be discussed in this case study. For simplicity, the example ignores steps like cooling and calibration which might be needed in a real process chain.

It is based on the results of the Cluster of Innovation "TurPro" [14] for tip repair of gas-turbine airfoils. A laser material deposition machine tool and a milling machine tool with an integrated laser-scanning device were used to repair the tip. This setup allowed to scan, mill and recontour on a single machine tool. A unity reference coordinate system was guaranteed by a specifically developed clamping system for turbine blades.

The activity diagram in Fig. 5 shows by repeating the four main process steps, laser scanning, milling, laser material deposition and recontouring, for each blade of the BLISK. One option would be to directly transfer this process chain to a BLISK and executing it on each blade. However, this process chain would be highly inefficient since a BLISK with n blades requires $2n + 1$ manual clamp and unclamp operations and $2n$ scan/milling tool changes.

Execution of a single process type for all blades of the BLISK at a time followed by switching to another machine tool shows an increased efficiency. This

Fig. 5 Naive adaption of blade process to BLISK

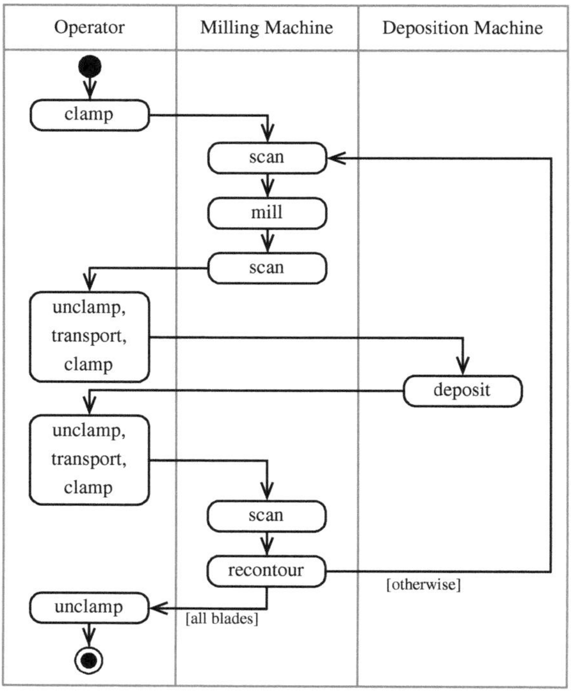

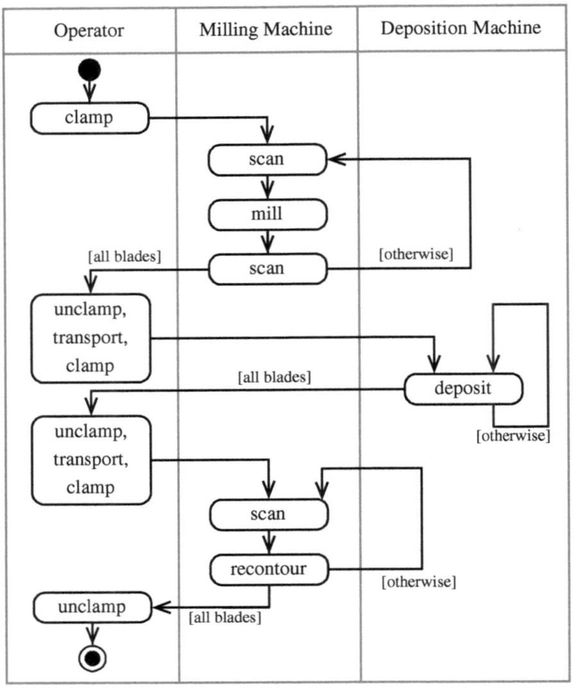

Fig. 6 Optimized process layout with less time consuming activities

approach as it is illustrated in Fig. 6 reduces the amount of manual clamp and unclamp operations to a total of just 3 and 2 scan/mill tool changes.

This shows that the process chain for repairing the tip of a single blade can not directly be reused to repair the tips of multiple blades on a BLISK. It is necessary to insert loops in order to repeat process steps individually. Instead of copying the whole function block net of the original process chain and manually adjusting it, the GPBM and its additional abstraction level is used. With this method each of the three main process steps for one blade can be modeled in a simplified way as a "Generic blade process step", as shown in Fig. 4.

The "Laser Scanning" function block requires the nominal blade geometry and the coordinate system of the current blade. On receiving the RUN event, this function block scans the blade and calculates a point cloud representing the worn blade geometry, upon finishing the FIN event will be triggered. Next is the "Analyzer" function block, which determines the deviations and the actual position of the blade relative to the nominal geometry. Afterwards the "Check Constraints Placeholder" function block is expected to check if the previously determined deviation is not exceeding the maximum tolerances for the actual process. Defining constraint checking by using a placeholder is required since the type of check may vary depending on the process, the machine tool or the used material. On a successful check the "Process Placeholder" function block and therewith the actual process will be triggered.

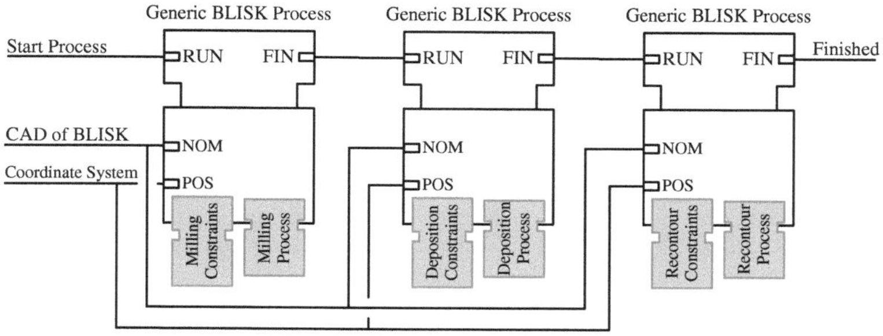

Fig. 7 BLISK tip repair process chain using instantiated generic function blocks

In Fig. 4 the generic blade process step is shown as a composite function block and instantiated three times to model the process chain for blade tip repair. Each of the steps is instantiated with one of the processes and its corresponding constraints function block.

The complete process chain for BLISK tip repair is shown in Fig. 7. By nesting the generic blade process into another GPBM, the previously mentioned loop for a BLISK is implemented. Therefore this block requires the nominal CAD model of the BLISK and its coordinate system as input. Either by user interaction, feature recognition or mathematical function, this function block will iterate over all blades of the BLISK, executing the nested generic blade process with the appropriate data. Identically to the generic single blade process step, this generic loop function block is used and instantiated with the actual milling, laser material deposition and recontouring function blocks and their corresponding constraints check function blocks.

The similarity between the modeled single blade and the BLISK process chain in Figs. 4 and 7 shows how process knowledge can be shared across different parts and thus confirms the potential of the GPBM.

Conclusion

This paper has proposed the new approach of "Generic Process Blocks for Manufacturing" (GPBM) by adding generics to IEC 61499. This concept enables programming of offline adaptive process-chains due to the characteristics of the function block standard and introduces another level of abstraction from which the user can benefit in multiple ways. Stored GPBMs can be reused in different scenarios and therefore help to preserve knowledge at a centralized place and by this simplifying the legibility of the model. Furthermore, process-related changes in the GPBM will automatically update and optimize all dependent processes without further user interaction.

References

1. TurPro: CAx-Framework—Fraunhofer-Innovationscluster TurPro. http://www.turpro.de/de/projekt/cax (2013). Accessed 28 April 2013
2. Shen, W., Wang, L., Hao, Q.: Agent-based distributed manufacturing process planning and scheduling: a state-of-the-art survey. IEEE Trans. Syst. Man Cybern. Part C Appl. Rev. **36**(4), 563–577 (2006)
3. Laguionie, R., Rauch, M., Hascoët, J.Y., et al.: Toolpaths programming in an intelligent step-nc manufacturing context. J. Machine Eng. **8**(1), 33–43 (2008)
4. Brecher, C., Vitr, M., Wolf, J.: Closed-loop capp/cam/cnc process chain based on step and step-nc inspection tasks. Int. J. Comput. Integr. Manuf. **19**(6), 570–580 (2006)
5. Rauch, M., Laguionie, R., Hascoet, J.Y., Suh, S.H.: An advanced step-nc controller for intelligent machining processes. Robot. Comput.-Integr. Manuf. **28**(3), 375–384 (2012)
6. International Electrotechnical Commission: Iec 61499-1: Function blocks—Part 1 architecture. International Electrotechnical Commission, Geneva (2005)
7. Xia, F., Wang, Z., Sun, Y.: Function block based design pattern for flexible manufacturing control system. In: IEEE International Conference on Systems, Man and Cybernetics 2003, vol. 5, pp. 4922–4927. IEEE (2003)
8. Minhat, M., Vyatkin, V., Xu, X., Wong, S., Al-Bayaa, Z.: A novel open cnc architecture based on step-nc data model and iec 61499 function blocks. Robot. Comput.-Integr. Manuf. **25**(3), 560–569 (2009)
9. Wang, X.V., Xu, X.W.: Dimp: an interoperable solution for software integration and product data exchange. Enterp. Inf. Syst. **6**(3), 291–314 (2012)
10. Vyatkin, V.: Iec 61499 as enabler of distributed and intelligent automation: state-of-the-art review. IEEE Trans. Industr. Inf. **7**(4), 768–781 (2011)
11. Brown, W.H., Malveau, R.C., Mowbray, T.J.: AntiPatterns: refactoring software, architectures, and projects in crisis. Wiley, New York (1998)
12. Vyatkin, V.: Iec 61499 Function blocks for embedded and distributed control systems design. ISA, USA (2011)
13. Gamma, E., Helm, R., Johnson, R., Vlissides, J.: Design patterns: abstraction and reuse of object-oriented design. Springer (2001)
14. Altmüller, S., Bergs, T., Börsch, W., Jelich, C., Klocke, F., Kriegl, B., Murtezaoglu, Y., Siebenwurst, C., Witty, M.: Beherrschung adaptiver prozessketten. In: Wettbewerbsfaktor Produktionstechnik, Aachener Perspektiven, pp. 405–422. Aachener Werkzeugmaschinen-Kolloquium, Shaker, Aachen (2011)

The Initial Analysis of Ethernet Bus for Monitoring HSM Process in Aerospace Industry

Piotr Szulewski

Abstract The monitoring of machining process seems to be the crucial for effectiveness and accuracy. The precise monitoring needs fast, efficient, easy and open communication standard for data interchange or data acquisition. The use a popular, inexpensive and ordinary network system instead complicated and valuable professional/industrial field-bus is really attractive. The paper presents some conception of the office Ethernet implementation as data transmission structure for monitoring. The basic, laboratory tests with different protocols and configuration were done. As a conclusion several rules were formulated to help potential users.

Keywords Ethernet · Network · Monitoring · Data acquisition · Information integration

Introduction

During the machining materials with high cutting speed several problem occurs. For example: chatter vibration in all axis, high temperature (tool and work piece), flexible cutting condition, work piece deformation, etc. [1]. All these parameters affects on work piece quality. Because it seems that the tool condition is the most crucial and determining factor for quality—the online tool condition monitoring is of great industrial interest. It is incredibly important, especially in aerospace industry, where expensive combination alloys are machining and quality is crucial. Therefore a sophisticated process monitoring system is strictly required to identify the actual cutting conditions. Usually, the basic monitor system for multi-modal

P. Szulewski (✉)
Faculty of Production Engineering, Warsaw University of Technology, Warsaw, Poland
e-mail: maxer@cim.pw.edu.pl

B. Denkena (ed.), *New Production Technologies in Aerospace Industry*,
Lecture Notes in Production Engineering, DOI: 10.1007/978-3-319-01964-2_22,

sensing consists of several nodes like: force sensor, vibration sensor, acoustic emission sensor, feed rate and cutting speed sensor, temperature sensor [2]. From the informatics point of view—plenty of generated data coming from many sensors should be transferred directly from their source to supervising system (computer). But how to find a good communication solution which will be really effective, open, scalable, inexpensive and easy to use?

Thesis

The main goal of realized experiment, was testing the influence of network topology and network load on data real delivery time. Proposed (working) thesis was: In case, when transmission in real time in not absolutely necessary the normal, traditional Ethernet should be fully used. It can be use as effective bus for process monitoring instead complicated and expensive industrial field buses. As a traditional Ethernet is characterized office installation and standard TCP/IP stack. All tests were done with artificial network transmission which was prepared to imitates the real communication between sensors and computer.

Background

In today's really competitive world, where market determines success and principles, all manufactures have realized that they absolutely need a "total automation" as their mantra to triumph. It should be stated that the factory automation has been lately developed worldwide into very large research area. In fact, it presently incorporates and analyze completely different knowledge disciplines including control, sensor, actuators, communication, real time adjustment, software programming, etc. However we can observe that the main impact of automation system (especially in machine tools) is focused on high effectiveness of process monitoring or control. The possibility of precise (close loop) supervisory system which integrates individual process characteristics will result in really higher productivity, machine tool effectiveness, better final product quality and expected near-zero breakdown [3]. The modern manufacturing systems (like machine tools, robots, technological equipment, assembling lines, etc.) are controlled by programmable numeric controllers (PLC, NC, CNC) or industrial computers. Sometimes the embedded, dedicated microcontrollers should be used to run control algorithm. The computation and processing of digital information is the basement for present control and monitoring. This situation is sometimes called as information cloud or information distribution. To realize the right communication between all nodes—the electronic data interchange system should be used. Many different network types have been promoted and recommended for use on a shop floor. Over the past 20 years several field bus network types were introduced in factory automation. But in fact, many years of standardization effort have completely failed to achieve one, good, universal, single standard. On the other side, there is Ethernet which has been around for many years and it is predominant in office computer systems with more than 85 % of the LAN market [4]. Cheap, easy to manage, compatible components for Ethernet-base networking are really attractive for everybody. Nowadays, a considerable number of factory and control automation applications adopted Ethernet as its industrial communication solutions [5]. Ethernet has been originally developed by Xerox and is known more than 40 years. But in fact, this standard was commercially introduced in early 1980. It is fully compatible with ISO/OSI regulations and belongs to the local area network (LAN). As a main advantages should be enumerated: compatibility, higher bandwidth, media access control, different physical media, reducing installation costs, increase reliability, improve management, practically no limit distance, reduction of electrical noise problems, provides electrical isolation, etc. Ethernet uses a communication concept called datagram to get messages across the network. The stream of data is divided into shorter pieces called frames. Every frame contains source and destination addresses and error-checking data so that any damaged data can be easy detected and re-transmitted. The data field in each package can contain up to 1500 bytes of user information. All over the world the Ethernet is qualified as a synonym of really open network standard which is define with attributes: interconnectivity, interoperability and interchangeability. It is also the most popular network standard in office and home environment. Mainly, the Ethernet is precisely described by norm signed as IEEE 802.3x where "x" indicates following standards. The main difference between standards respects the throughput, topology, distance range, medium type. Historically Ethernet technology was established as a standard for office communication to provide effective communications between the different nodes like PCs, printers, data servers etc. In was thus no surprise that industry sector was looking carefully to Ethernet for a really low cost networking solution. The main problem in simple adoption this conception into industrial requirements was the non shared and non dedicated network bandwidth. The transmission collision could occur at any time and it was impossible to predict message delays. The (office) Ethernet is in fact non deterministic. For these reasons, as a result for the constant development, the special variety called industrial Ethernet has been created and implemented into automation systems [6]. The main differences between Industrial and office Ethernet are: support for real-time application, redundancy, accommodate multiple service quality levels, time synchronization, fault tolerance, ready for work with high humidity (condensing), strong vibration, hot temperature, corrosive environments, etc. Presently, many products dedicated for industry are available on the market. Most popular could be enumerated: EtherNet/IP, Profinet, ModbusTCP, Powerlink, EtherCat. Unfortunately these standards are completely incompatible eachother. On the other hand they are extremely expensive due to the necessity of purchase only dedicated equipment (hardware) and specialized software.

Of course it is possible to use a special, dedicated equipment for translation data stream from one standard to another. As an example of such a transfer method the

AnyBus bridge can be use. But, translation process needs time (effective throughput decreases) and is not cheap. So, the traditional Ethernet technology is still needed, useful and attractive for industrial use—especially if it will be little tailored for monitoring processes. It should be emphasized that traditional Ethernet makes process data and diagnostic functions device-independent and all equipment in a plant should be linked with just one bus technology (fast and inexpensive) [7]. On the other hand, there is a significant difference between industrial and office operational reliability. The equipment MTBF-value may be important for professional installation, but for laboratory use is not pivotal.

Installation—Hardware

The laboratory test structure installation is presented in Fig. 1. It consists of: notebook (PC 3) with operational system Win XP (32-bit, Hewlett-Packard Compaq 6710 s), network Ethernet terminal (Beckhoff, modul BK9050—Bus Couplers connect Ethernet with the modular, extendable electronic terminal blocks—KL2408.), stationary personal computer with Linux Slackware (ver. 13), Ethernet switch Cisco (Catalyst 2900XL with firmware C2900XL-c3h2 s-ms.120-5.WC3b) and power supply NDN DF (24VDC/5A). The TwinCat (ver. 2.11.1553, Beckhoff) software is installed and configured on the notebook. This computer works as a ordinary PLC under control of software "PLC Control" made by Beckhoff. For the maximum computer performance all not necessary system functions were in switch off mode to move computational power only into useful tasks. The stationary computer works like a substitute of simple data acquisition server which collects measured data from different sensors. Three different network structures has been chosen to check the network traffic for possible process monitoring configuration during the real machining in HSM. All network components are connect using unshielded twisted pair cables (cat. 5e) ready for 100 Mbit/s bandwidth. The standard (in fact "stationary") packet delivery time was checked using the "ping" tool. The received result was not surprised—1 ms, which is absolutely proper response time for such structure.

Installation—Diagnostic Software

As a measurement device some sophisticated software tools were used. On the stationary computer the "Wireshark" software was installed. It is a specialized data monitor and performance meter for network application. Also, for general traffic monitor the "iptraf" tool was applied. The graphic presentation of indicated data transmission were done using the standard windows network monitor tool. The intercepted network data were analyzing using the Excell spreadsheet and presented in graphic form for useful perception.

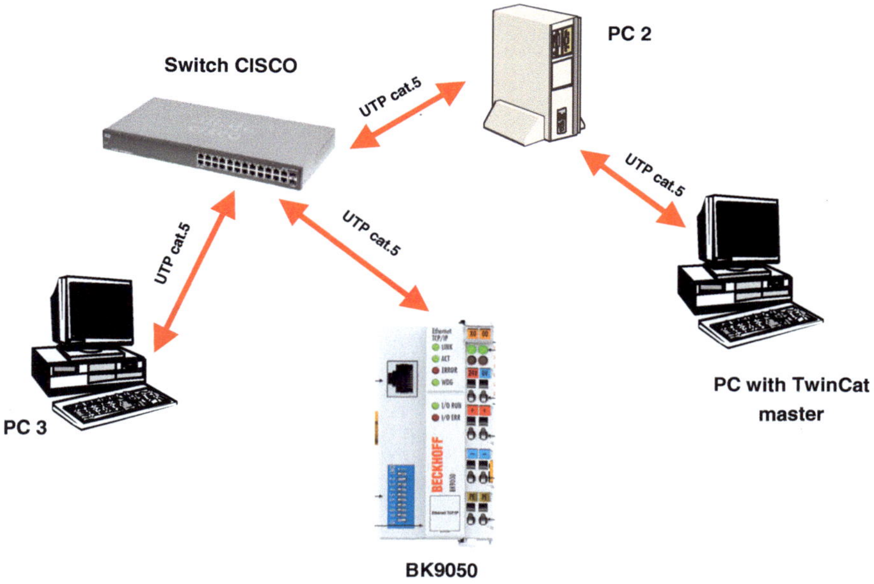

Fig. 1 The main installation structure

Methodology

The really important difference between industrial and office application is the data sent frame size [8]. The typical industrial communication data size is rather short and usually does not exceed 1–8 bytes. For the test—64 bytes was approved. This is the minimum Ethernet frame size. This is no profitable situation for real data acquisition but test should be done in the worst applicable case. The Fig. 2 presents the first 10 captured packets during the initial conversation between PC2 and BK9050. It is significant the RTT is really short and not exceeds the value 7.8 ms. The data was captured using the "wireshark" network tool. For example, the main computer collects information from several vibration sensors mounted on machine tools—data are transferred together in one packet. In fact the protocol reduces slightly the data transmission efficiency. As a important transmission parameter the round trip time (RTT) was accepted. This concept identify the time for short data transmission from one network node to another (PLC to PC or sensor to PC). It is described by simple formula: $T_1–T_0 = RTT$ (where T_1—response receive time indicated by "Wireshark", T_0—demand time indicated by "Wireshark"). Transmission times were observed and reported for TCP protocol and UDP protocol separately. Differences between these protocols are considerable and should be noticed. UDP is connectionless protocol. It uses a simple transmission model with a minimum of protocol mechanism. It is suitable for purposes where error checking and correction are either not necessary or performed in the application,

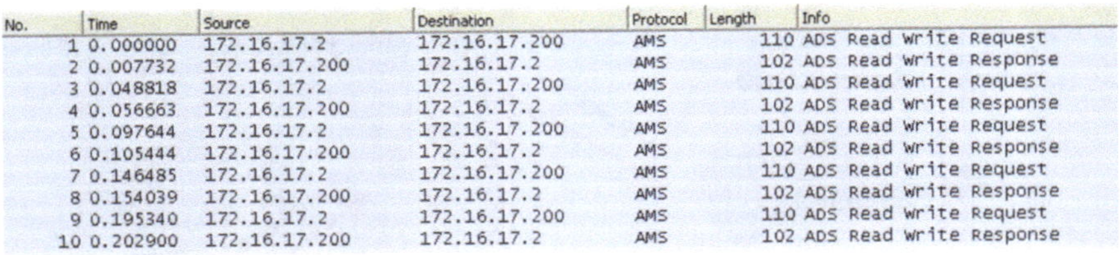

No.	Time	Source	Destination	Protocol	Length	Info
1	0.000000	172.16.17.2	172.16.17.200	AMS	110	ADS Read Write Request
2	0.007732	172.16.17.200	172.16.17.2	AMS	102	ADS Read Write Response
3	0.048818	172.16.17.2	172.16.17.200	AMS	110	ADS Read Write Request
4	0.056663	172.16.17.200	172.16.17.2	AMS	102	ADS Read Write Response
5	0.097644	172.16.17.2	172.16.17.200	AMS	110	ADS Read Write Request
6	0.105444	172.16.17.200	172.16.17.2	AMS	102	ADS Read Write Response
7	0.146485	172.16.17.2	172.16.17.200	AMS	110	ADS Read Write Request
8	0.154039	172.16.17.200	172.16.17.2	AMS	102	ADS Read Write Response
9	0.195340	172.16.17.2	172.16.17.200	AMS	110	ADS Read Write Request
10	0.202900	172.16.17.200	172.16.17.2	AMS	102	ADS Read Write Response

Fig. 2 The RTT for initial conversation—"Wireshark" print screen

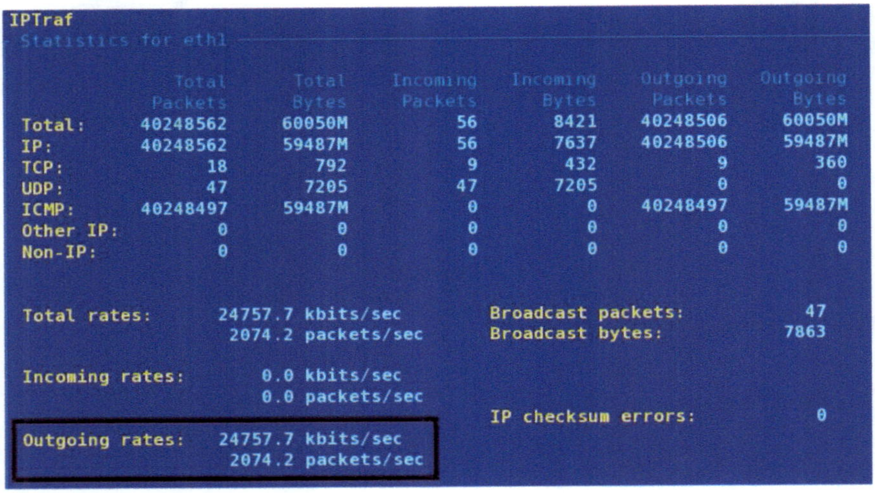

IPTraf

Statistics for eth1

	Total Packets	Total Bytes	Incoming Packets	Incoming Bytes	Outgoing Packets	Outgoing Bytes
Total:	40248562	60050M	56	8421	40248506	60050M
IP:	40248562	59487M	56	7637	40248506	59487M
TCP:	18	792	9	432	9	360
UDP:	47	7205	47	7205	0	0
ICMP:	40248497	59487M	0	0	40248497	59487M
Other IP:	0	0	0	0	0	0
Non-IP:	0	0	0	0	0	0

Total rates: 24757.7 kbits/sec
2074.2 packets/sec

Broadcast packets: 47
Broadcast bytes: 7863

Incoming rates: 0.0 kbits/sec
0.0 packets/sec

IP checksum errors: 0

Outgoing rates: 24757.7 kbits/sec
2074.2 packets/sec

Fig. 3 The broadcast storm in network—"iptraf" print screen

avoiding the overhead of such processing at the network interface level. On the other hand, TCP is connection-oriented protocol. It means that TCP provides reliable, ordered, error-checked delivery of a data stream. TCP is optimized for accurate delivery rather than timely delivery.

Research Work

The data broadcast storm is presented in Fig. 3. It occurs when network is wrong configured, viruses are working, DoS attack, etc. The stationary PC 2 has sent 2100 ICMP packets per second. It is estimated to the 25 Mb/s network throughput. In such a worst situation the real useful transmission between BK9050 and PC3 is stable and acceptable. The outgoing rates 24757.7 kbit/s is marked by dark rectangle in figure. During the research work, the several scenarios were prepared, simulated, tested and analyzed. As an example, some main tasks will be enumerated to show the complexity of transmission problems: (a) The communication in simple (peer-to peer), network structure, (b) The communication in complicated architecture when the switch works as a transmission mediator, (c) The difference in data encapsulation in TCP and UDP protocol, (d) The double transmission in data-pair on demand, (e) The communication behavior in separate collision domain, (f) The communication in broadcast and unicast mode, (g) The simulation of packet collision, (h) The software redundancy for stability of data transmission, (i) The virtual private network segmentation as solution for reducing the traffic storm, (j) The implementation of QoS (Quality of Service) at the media access control (MAC) level to divide data stream into referred priority to minimize the time delivery. The Fig. 4 indicates the graphic presentation of network use during the transmission storm. The average network load was 60 Mb/s. It can be easy observed when packet collision appeared—so utilization is significantly decreased. The reason for storm was false Vlan configuration. Simple division into separate independent Vlan solved that problem completely. The Fig. 5 presents the RTT for transmission of 10.000 packet (5000 pairs) between PC3 and BK9050 using TCP protocool. answers The minimum RTT was 7.63 ms and the maximum 23.49 ms. The standard deviation was 1.7 ms. It seems that use of TCP in that case is acceptable.

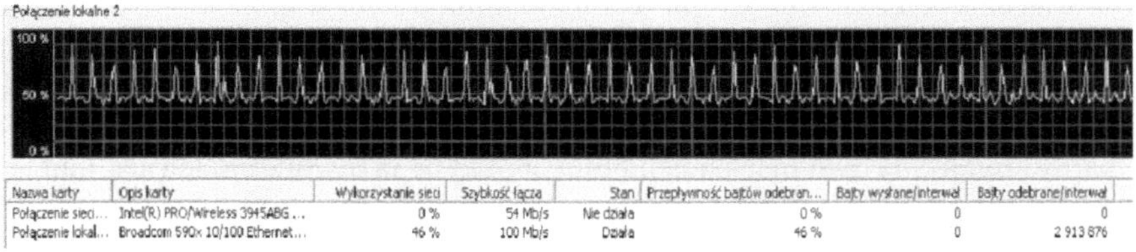

Fig. 4 The network throughput in storm—“network monitor” print screen

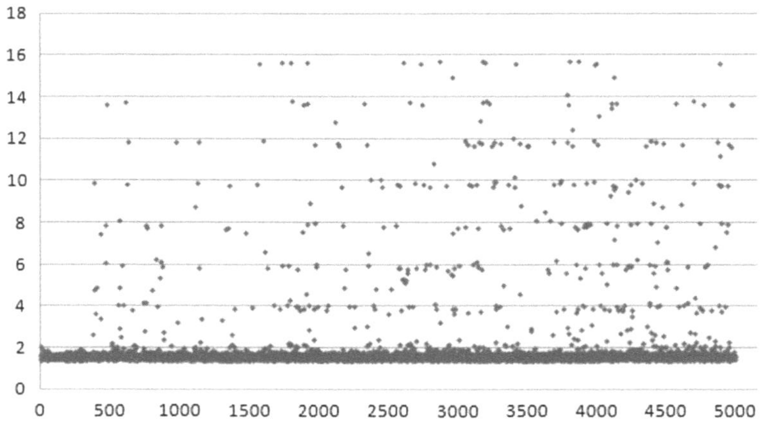

Fig. 5 The RTT (min 7.63 ms, max. 23.49 ms) for 10,000 transmissions by TCP protocol

Conclusion of Research

According to the achieved results of realized tests and research work, some general, useful conclusions should be formulated and presented. They will be beneficial for constructing the modern, effective acquisition system for machine tool and process monitoring. Please note that:

- Some basic security regulation should be introduced into Ethernet network structure to prevent traffic jam (data storm) and to acquire the solid stability of transmission.
- The conception and Ethernet components make possible to achieve real scalability of network architecture and topology. Also, the implementation and configuration process is shorten to minimum time and obligate effort. Well known arrangement and intuitive interface reduces the required training for personnel.
- The multitude of different available protocols help to find the best (may be tailor made) solution for every transmission task.
- If the most important for slave-master communication is reliability without necessity of short time delivery, the TCP/IP protocol should be recommended. On the other hand, if really important is short time and more predictive behavior of transmission, the UDP/IP protocol should be rather use.
- In critical situation when effectiveness of transmission is the main goal—the software redundancy should be implemented. For example, the simple STP will be enough for many application. Every producer of the network Ethernet hardware has his own mechanism but it should be in a work mode (just find and run it).
- It is strictly recommended not to engaged the main computer into unimportant tasks during the data acquisition process. All, not necessary tasks (www, mail, etc.) should be closed or cancelled. If it is not possible to restrict the communication with other computers, the software priority of network datagrams should be implemented.
- In case of large, complex data acquisition network, to reduce the possibility of unexpected attacks (DoS-Denial of Service, broadcast storm, data overload or useful bandwidth reduction) the virtual structures (v-lan) should be created. It is the best solution to separate traffic between different measurement structures.

Referring to the thesis proposed at the beginning of the paper, it is justify that if some rules and limits are accept, the performance of traditional Ethernet is absolutely applicable for data transmission bus to enhance flexibility and reduce final costs. But from the practical point of view, the test with real data coming from sensors is strictly necessary to check the stability and network load. Such a work will be done in nearest future. An extra advantage of Ethernet implementation is simplicity cooperation with CNC controllers. Many current CNC are equipped with Ethernet interface as a standard so double-way communication with monitoring equipment will be easy. It is important that monitoring mechanism should has a real influence on monitored process to change wrong parameters or correct variables.

Acknowledgements Financial support of Structural Funds in the Operational Programme—Innovative Economy (IE OP) financed from the European Regional Development Fund—Project "Modern material technologies in aerospace industry", Nr POIG.01.01.02-00-015/08-00 is gratefully acknowledged.

References

1. List, G., Sutter, G., Bi, X.F.: Investigation of tool wear in high speed machining by using a ballistic set-up. Wear **267**(9–10), 1673–1679 (2009)
2. Zeng, H., Thoe, T.B., Li, X.: Multi-modal sensing for machine health monitoring in high speed machining. In: 2006 IEEE International Conference on Industrial Informatics, http://ieeexplore.ieee.org/stamp/stamp.jsp?arnumber=04053566
3. Ridwan, F., Xu, X.: Advanced CNC system with in-process feed-rate optimisation. In: 2013, Robotics and Computer-Integrated Manufacturing, vol. 29, pp. 12–20. Elsevier
4. Xavier Desforges, X., Habbadi, A., Archimède, B.: Design methodology for smart actuator services for machine tool and machining control and monitoring. Robotics Comput. Integr. Manufact. **27**(6), 963–976 (2011)
5. Rostan, M.: Industrial ethernet technologies overview. EtherCAT Technology Group (2011)
6. Ferraz, F., Coelho, R.T.: Data acquisition and monitoring in machine tools with CNC of open architecture using internet. Adv. Manufact. Technol. **26**, 90–97 (2005)
7. Samaranayake, L., Leksell, M.: Ethernet ready sensor acutator module for distributed control application. Eurocon 2005, Belgrade, Serbia and Montenegro, pp. 342–345. University of Belgrade (2005)
8. Tan, K.K., Huang, S.N., LEE, T.H.: Distributed fault detection in industrial system based on sensor wireless network. Comput. Stand. Interface **31**, 573–578 (2009)

Producing Better Turbines by Using Process Monitoring and Documentation Technologies

Jan Brinkhaus, Martin Eckstein and Joachim Imiela

Introduction

Modern aircrafts are designed under aspects of passenger safety, weight as well as running, purchasing and maintenance costs. These aspects conflict with each other at several points. Especially limiting the aircraft costs and weight conflicts with the most important aspect—the passenger safety. KOMET Brinkhaus and MTU Munich established in compliance with FAA rules a whole production strategy which allows a reduction of safety factors in the design of turbines without reducing the aircraft safety. The new turbines are light weighted and thus fuel efficient. The primary enabler was a strict real time process monitoring with complete process documentation while producing safety critical elements.

Today propulsion systems for planes have to meet increasingly more challenging requirements in terms of performance, cost-effectiveness, operational safety and environmental compatibility. To be able to compete in the international marketplace, engine manufacturers must develop innovative solutions for the design, manufacturing and quality assurance of highly stressed jet-engine components to satisfy the requirements of airlines and safety regulations [1–4].

Jet-engines clearly differ from other propulsion systems used in general mechanical engineering and vehicle construction above all by the materials used and the extreme stresses occurring in operation. Highly-stressed jet-engine

J. Brinkhaus (✉) · J. Imiela
KOMET Group, Besigheim, Germany
e-mail: Jan.Brinkhaus@kometgroup.com

M. Eckstein
MTU, Munich, Germany

J. Imiela
University of Applied Sciences and Arts in Hanover, Hanover, Germany

B. Denkena (ed.), *New Production Technologies in Aerospace Industry*,
Lecture Notes in Production Engineering, DOI: 10.1007/978-3-319-01964-2_23,

components have to meet the highest requirements in terms of reliability and component service life. Therefore very specific solutions are required [2, 5–7].

Bolt holes in rotating turbine and compressor disks belong to the most highly-stressed geometric features of jet-engines. Their in-flight-failure may result in the loss of the aircraft and endanger human lives. For this reason, the quality and reliability of hole making are of utmost importance. For manufacturers of jet-engine components it is important to assess the quality of these holes at an early stage in the manufacture of the product and not as late as at final inspection. In the event of apparent deviations, it must be possible to interfere with the manufacturing process in time, to make sure that component properties are within the specified limits, defined at the design stage and validated and certified in engine test runs [1, 6].

Among other manufacturing methods, the machining of holes in jet-engine parts is especially sensitive, because a lot of them are situated in highly stressed areas. In the past, improper hole-making has already caused damage or inflight-shutdown of engines. The most popular of these engine failures is the so called "Pensacola-Incident", which caused the death of two passengers in 1996 [8]. The reason for this undiscovered failure was the misjudgment of a tool breakage during pre-drilling of a compressor disk. Figure 1 gives an impression of the enormous energy which was set free.

The tool breakage itself was caused by a lack of coolant during drilling. The component did not fail immediately but after approximately 10 years in use. This was caused by debris of the broken tool which remained in the component and which was not detected by inspections [8].

All of the above aspects suggest the production of mechanically stable and heavy turbine components. Nevertheless it has to be considered, that constantly increasing standards for environmental protection e.g. for noise, NOx- and CO2 emissions do not automatically support the increase in robustness of aero-engines. These limitations impose additional pressure on engine designers to reduce weight, improve aerodynamics, increase process temperatures and enhance operational

Fig. 1 Pensacola-incident

loads. This challenge can only be mastered with an improved control of the quality of the manufacturing processes.

The purpose of monitoring the production processes is to reliably identify deviations or anomalies during machining immediately and take appropriate action [9, 10]. This protects the machine-tool and/or work piece against damage [11, 12], saves related costs, secures or even improves product quality [13–15] and enables unmanned production.

Beneath the general argument of better process control, a second motivation for monitoring has gained importance for the safe life concept of aero-engine design. FAA-AC 33.3 [16] describes, that within the design process of highly stressed aero-engine parts—in addition to stress calculation—a special risk assessment has to be performed [17]. This is called damage tolerance assessment. If the result of this assessment does not provide enough evidence that the target life of a component or it's life-limiting features can be achieved, a risk mitigation is necessary. By choosing creditable means to improve production quality, the output of the risk assessment can be updated. These credits may save a product redesign, a change in material or additional in-service inspections [17]. Creditable means are for example the introduction of a post finish process, an advanced component inspection or statistical data for a high process capability. By far, the highest credit is granted for process monitoring.

Monitoring the Production

MTU and KOMET Brinkhaus have applied a very strict strategic combination of process monitoring and process documentation for the production of holes at the MTU plant in Munich. The KOMET product ToolScope was enhanced and embedded into a consequent strategy that targets a complete production process control. The whole strategy was developed and implemented during a 5-years R&D effort.

As prerequisite the MTU started a research project several years ago, which targeted the basics of the needed monitoring strategies. It was evaluated, which monitoring strategies where needed to fulfill the FAA rules as best as possible. Therefore it was determined, which effects disturb a constant drilling and how they could be measured in which sensor data. Details of these aspects will be published in other publications.

Based on this research, a very deep strategy of monitoring the production process was developed and implemented. Parts of this strategy are the following aspects:

(1) Monitoring the coolant flow and pressure:
By monitoring the coolant flow and pressure, two big requirements for a constant drill process are assured. The correct flow is a precondition for a

defined temperature at the tool tip as well as the correct inprocess-removal of machined material out of the hole.

(2) Monitoring the spindle and feed axis power:
Defined conditions for spindle and feed axis power are also preconditions for a constant drill process. They guarantee primarily a constant force and temperature foot print at the tool tip. Together with the simple values of the powers, some attributes of the signals are extracted and monitored separately. These attributes were found and validated in the mentioned MTU research program.

(3) Assuring that the machine personnel does not disable or work around the process monitoring devices:
During production ramp up it was seen, that the machine personnel has also to be monitored not to modify the process. Because the monitoring of many production values was very tight, it was primarily needed to assure, that the machine doesn't produce, if the monitoring system is brought in a condition where it would not work. This was done by a close handshake between PLC and the monitoring system.

(4) Documenting information like part numbers, NC program versions, used tool names etc.:
To let the monitoring system do the main part of the documentation, it was needed to transfer a big amount of data from the production machine to the monitoring system. To avoid that the user has to do too much manual input at the machine before each process, as much information as possible was hard coded as variables within the NC program. Some parts of the information are then asked from the user during production. Before each single hole, all data is transferred by special NC cycles from the NC of the machine to the monitoring system. For this, untypical features of a process monitoring system were needed. An example is that the monitoring system needed the ability to exchange strings (like "program name") with the control.

(5) Creating measurement protocols and preserve the relevant part information's and sensor data for each hole:
Much work was spend in a good and practicable scheme of documenting the monitoring results. It was seen during tests, that many schemes exist which have for each their pro's and con's. The first documentation variants were implemented as Excel readable ASCII files, together with screenshots of the monitoring. One step forward was the generation of nicely formatted PDFs which gave all needed information in a printable paper. A next step will be the integration of the monitoring system into a central measuring data store of the MTU. This store—of course—has special data formats too. Many of the interim results are still present in the current implementation of the production monitoring, because every of the mentioned documentation formats has its own advantages.

(6) Archiving this information safely at external long time data stores:
The aspect of long time data storages was a big topic at the end of the project. The monitoring system was extended by the ability of handing out its protocol

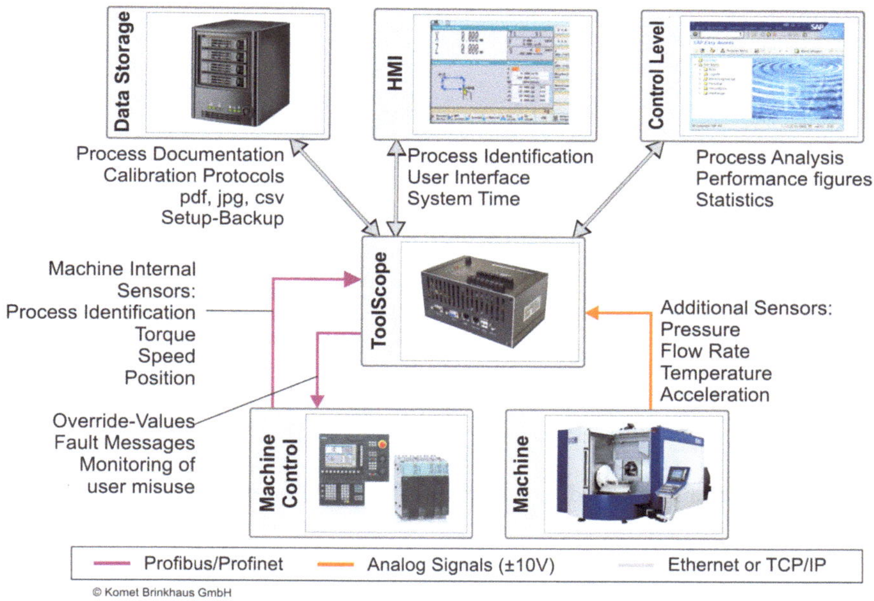

Fig. 2 Setup of the described strict monitoring and documentation engine

data within windows networks. For getting a clean data flow design, handshakes had to be established between the control, the monitoring system and the documentation system. To achieve a clear network design, the monitoring system offers two network ports, which can be configured for separate IP address ranges. The system can thus be connected to the machine internal network and an external factory network, without needing a direct connection of both networks with a switch or router.

The structure of the monitoring engine is shown in Fig. 2.

Outlook

MTU Munich and KOMET Brinkhaus could achieve the—apparent currently highest—credit value from the FAA for hole making by applying the above described strategy. The whole system is currently implemented on four CNC machines in Munich and two CNC machines at a supplier plant in Israel. This structure will perhaps be extended in short time by four additional machines in Munich. Current works are dealing with the user interface, the exchange of monitoring limits between machines and the integration into other MTU data stores.

Authors

Dr. Jan Brinkhaus founded a process monitoring company which is now part of the KOMET group. He is CEO of KOMET Brinkhaus, specialist for process monitoring and the integration of monitoring systems in Siemens and Heidenhain controls.

Dipl.-Ing. Martin Eckstein managed several MTU technology projects process monitoring, process control and coolant application. He is auditor for the production of security of highly-stressed turbine components. The audits include the MTU facility in Munich as well as suppliers all over the world.

Prof.-Dr. Dipl.-Kfm. Joachim Imiela managed several process and condition monitoring projects at the Leibniz University Hanover. He is currently professor at the University of Applied Sciences and Arts in Hanover and manages the world wide costumer consulting at KOMET Brinkhaus.

References

1. Adam, P. Fertigungsverfahren von Turboflugtriebwerken, 1st issue, Birkhäuser Verlag (1998)
2. Bräunling, W.J.G. Flugzeugtriebwerke, Springer-Verlag, Berlin (2009)
3. Burin, J.: Down time. AeroSafety World. Issue February 2012, pp 29–32 (2012)
4. Gunston, B.: Jane's Aero Engines, Jane's Information Group, ISBN 0710614055, 9780710614056 (1996)
5. Buergel, R.: Handbuch Hochtemperaturwerkstoffe, 3rd Issue, Vieweg-Verlag, Wiesbaden (2006)
6. Rossmann, A.: Die Sicherheit von Turbo-Flugtriebwerken-Band 3, Turbo Consult, Karlsfeld (2006)
7. Sims, C.T., Stolofff, N.S., Hagel, N.S.: Superalloys II, Wiley, New York (1986)
8. N.N.: Aircraft Accident—Uncontained Engine Failure Delta Airlines flight 1288. National Transportation Safety Board, Washington, D. C. (1996)
9. Kluft, W.: Zerspanprozess-Überwachung und mehr. Werkstattstechnik, **91**, 357–359 (2001)
10. Golz, H.U., Schillo, E., Wolf, A., Kaufeld, M., Sprengel, P., Johannsen, P., Heinek, D.: Bewertung von Werkzeugüberwachungs-systemen aus der Sicht der Anwender. VDI Berichte 1179, Düsseldorf, 309 ff (1995)
11. Brinkhaus, J.: Statistische Verfahren zur selbstlernenden Überwachung spanender Bearbeitungen in Werkzeugmaschinen; Promotion zum Dr.-Ing., ISBN 978-3-941416-31-4, PZH-Verlag, Hannover (2009)
12. Kaever, M.: Steuerungsintegrierte Prozessüberwachung bei spanender Bearbeitung, Dissertation, RWTH Aachen (2004)
13. Adam, P., Eckstein, M.: Zeitfestigkeit und Randschicht-Eigenschaften durch Zerspanung bei Nickellegierungen. Materialwissenschaft u. Werkstofftechnik, **27**, 272–279 (1992)
14. Brinkhaus, J.: Qualitätssicherung in der Fertigung hochbelasteter Bauteile; Proceedings: Sichere und zuverlässige Fertigung durch Prozessüberwachung 2009, Verlag WZL Forum GmbH, Aachen (2009)

15. Eckstein, M.: Process Monitoring-ein innovativer Ansatz zur Steigerung der Zuverlässigkeit höchstbelasteter Bauteile der Fluggasturbine, Deutscher Luft- u. Raumfahrtkongress, Jahrbuch (2004)
16. N.N.: DIN EN ISO 3059-Zerstörungsfreie Prüfung—Eindringprüfung und Magnetpulverprüfung—Betrachtungsbedingungen, Beuth Verlag GmbH (2013)
17. N.N.: FAA AC33-2B, FAA Airworthiness Directives and Advisory Circulars, Appendix, US Department of Transportation. http://www.faa.gov (2006)

From Fuzzy Maintenance, Repair and Overhaul Data to Reliable Capacity Planning

Steffen C. Eickemeyer, Simon Steinkamp, Bernhardt Schuster and Sebastian Schäfer

Abstract To increase the logistical performance of regeneration processes, the Institute of Production Systems and Logistics has developed methods for efficient capacity planning in spite of fuzzy load information. The methods were validated using real data and an analysis of the methodology for its sociocultural conformity and adaptability in a global company are being carried out.

Introduction

The topic of capacity planning is a partial project (D1) of the Collaborative Research Centres (CRC) 871 "Regeneration of Complex Investment Goods". Regeneration is the restoration, respective of the condition, of a complex investment good. Regeneration in this context is also known as "Maintenance, Repair and Overhaul" (MRO). Complex investment goods, in the sense of this CRC, are investment goods in which a high number of components stand in diverse,

S. C. Eickemeyer (✉)
Scientific Assistant in the Production Management Research Group, Institute of Production Systems and Logistics, Leibniz University of Hannover, Hannover, Germany
e-mail: eickemeyer@ifa.uni-hannover.de

S. Steinkamp
Temporary Scientific Assistant at the Institute of Production Systems and Logistics (IFA), Production Engineering Center Hannover (PZH), Leibniz University of Hannover, Hannover, Germany

B. Schuster
Director of Engineering and Technical Support at MTU, Maintenance Canada Ltd, Hannover, Germany

S. Schäfer
Production Planning Senior Manager at MTU, Maintenance Hannover GmbH, Hannover, Germany

B. Denkena (ed.), *New Production Technologies in Aerospace Industry*,
Lecture Notes in Production Engineering, DOI: 10.1007/978-3-319-01964-2_24,

functional relations. For example, complex investment goods are machine tools, wind energy equipment, aircraft engines, diesel locomotives, or stationary gas turbines [1].

Motivation and Bases

Creating value through the regeneration of goods is only achieved when the revenues earned through additional service life exceed the costs of all the regeneration work required [2]. However, planning regeneration maintenance suffers from uncertainties about the future (Fig. 1) [3]. Because of their complexity, not all repairs to goods can be precisely planned in advance. Furthermore, there is uncertainty regarding the necessity for the replacement of excessively damaged goods. The only information available is vague, which leads to challenges when planning the regeneration capacity [4]. Consequently, in order to guarantee efficient planning of capacities [5] with high on-time delivery and short turnaround times, a forecast of the regeneration work required for complex capital goods needs to be as precise as possible [6]. Combining a workload forecast with capacity planning in the field of complex goods for regeneration has hardly been studied scientifically and there are no approaches relevant to practice [5, 7]. That is why a method for capacity planning and adjustment for regeneration processes has been developed, which allows the maximization of target parameters despite fuzzy information. The entire project was designated for a four-year duration and will be completed in December of 2013 [8].

Generic Process Model

In the first step, a generic process model of regeneration regarding the corresponding information localization was developed. On one hand, the required as well as the available data and information in the corresponding process steps have been identified. On the other hand, information that already contributes to the

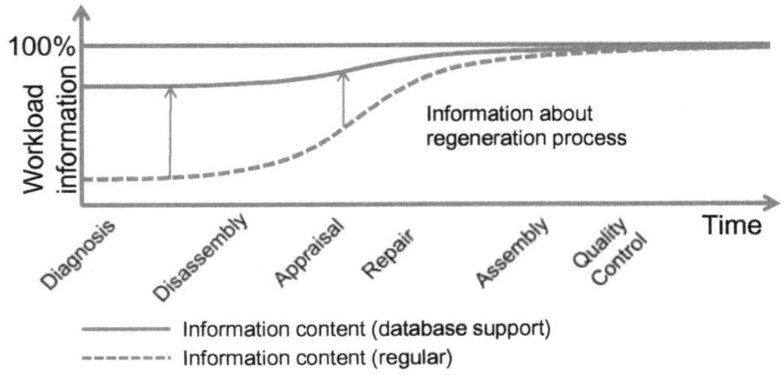

Fig. 1 Increasing accuracy of information in the Regeneration process [1]

localization or determination of specific damages was considered in early phases of regeneration and therefore gives considerable input for prompt capacity control. It is assumed that at a later point of regeneration, concrete information replaces information from earlier phases. In addition, the usage phase before the regeneration process was scrutinized concerning exploitable information [1].

Damage Library and Data Mining

The starting point for developing forecasts was a damage library in order to concretize information for capacity planning [1, 8, 9]. Within the data mining project, the damage library is used as a database. Bayesian Networks (Fig. 2) were identified as the most suitable method to forecast expenditures for regenerating complex investment goods despite uncertain load information [8, 9].

Whereas on one hand the information relating to the actual status of the components is constantly increasing, on the other hand we see a decrease in the capacity coordination options. Accordingly, capacity planning is carried out for different time periods; we distinguish between long-, medium-, and short-term planning. A Bayesian Network was developed for each planning horizon that takes into account the subdivisions. Implementing Bayesian Networks in actual test cases demonstrated that the generated forecasts for different levels could be positively evaluated (Table 1) [8, 9]. The forecasts of the Bayesian Networks have a distinct advantage over the rigid comparative values of the industrial partner, especially on the long- and short-term planning levels because current environmental influences are taken into account quicker.

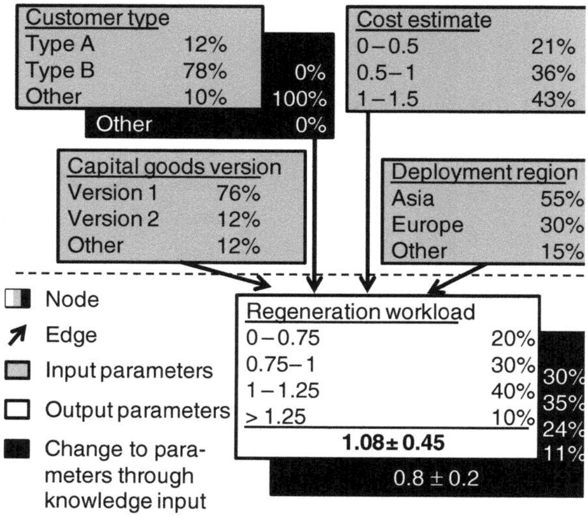

Fig. 2 Example of a Bayesian Network

Table 1 Assessment and validation results of Bayesian Networks (BN) and planned values of Industrial Partner (IP)

Planning	Long-term planning		Medium-term planning		Short-term planning	
Forecast	BN	IP	BN	IP	BN	IP
Absolute deviation [h]	540	840	570	600	174	295
Standard deviation [h]	420	720	390	480	124	173
More accurate forecast (%)	63.8	31.0	58.6	39.7	91.7	8.3

The networks allow valid statements regarding the anticipated regeneration expenditures to be drawn and can thus provide considerable support in planning capacities. In addition, the forecasts can make previous estimates about regeneration expenditures more precise [8, 9].

Capacity Planning and Results

Using the forecasts as a guide, three mathematical models were developed to determine the optimum labor force. The Long Term Model is a linear model, which determines the optimum labor force for each month. In order to plan appropriately for each month, the combination of in house and outside vendor labor equals or exceeds the forecasted labor hours obtained from the Bayesian Networks. The model maximized the target parameter, which in this case was profit. Utilizing IBM's ILOG CPLEX in conjunction with Microsoft Excel, a long-term plant capacity model was created and the results are shown below in Fig. 3.

The output data shows the optimal number of in house and outside vendor labor hours for each month in order to maximize profit over the course of a year. After the results were calculated, they were tested in real time with an industry partner.

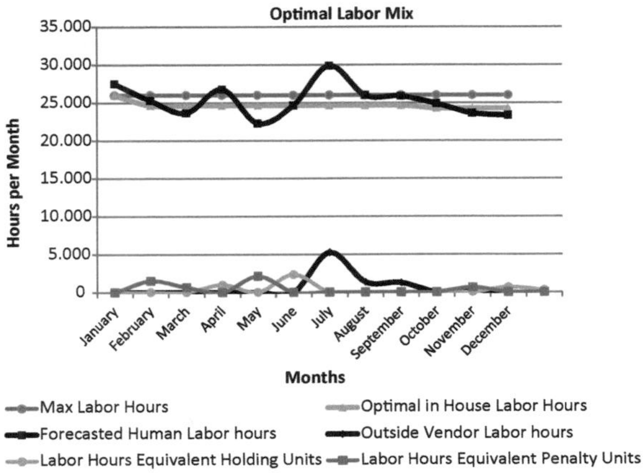

Fig. 3 The optimal labor mix for each month

Validation and Outlook

At the moment, logistical performance parameters such as the throughput times and the adherence to schedules of a regeneration area still need to be modelled and established in simulation experiments so they can be validated using real data with a cooperating partner. As the results of the past validations are very promising, it has to be determined whether these results are transferable to other applications and industrial plants in different countries with different products and cultures. To accomplish this goal, a second evaluation is taking place in a subsidiary of the cooperating partner in a different market environment.

Numerous studies show that the national cultures of a large number of nations have an impact on the transferability of capacity planning with regard to the effects on economics. For instance, both the study of Geert Hofstede [10] and the later approached GLOBE (Global Leadership and Organizational Effectiveness Research Program) study [11] identify cultural dimensions which are differently shaped in cultural areas and which are able to explain differing leadership concepts or the quality awareness of employees [12].

In addition to the national culture, both the size and age of a plant can influence its receptiveness to changes. For example, employees at a plant that is part of a new acquisition will probably react more suspiciously towards new processes and changes than the employees of a newly founded joint venture with a local company [13].

Although there are efforts to establish process standards in multinational companies to provide the same quality and products from every plant, there are always variances, especially in the MRO business of complex capital goods. They can be caused by a various number of reasons such as differences in plant size, products, customer composition or repair capacities. All of these factors have to be taken into account when transferring the regeneration workload forecast methodology and the functional model for capacity planning. They influence input parameters as well as considered limitations and bottlenecks. At the same time it is important to achieve the objective with as little effort as possible.

Acknowledgments The authors would like to thank the DFG research organization for providing funding for this research project within the scope of the CRC 871 program. The authors would also like to thank the Graduiertenakademie of the Leibniz Universität Hannover for providing funding for the validation of the theoretical results in the global industry.

References

1. Eickemeyer, S.C., Nyhuis, P.: Capacity planning and coordination with fuzzy load information. Bus. Rev. **16**(1), 259–264 (2010)
2. Freund, C.: Die Instandhaltung im Wandel. In: Schenk, M. (ed.) Instandhaltung technischer Systeme: Methoden und Werkzeuge zur Gewährleistung eines sicheren und wirtschaftlichen Anlagenbetriebes, 1st edn, p. 17. Springer, Berlin (2010)

3. Reményi, C., Staudacher, S.: Systematic simulation based approach for the identification and implementation of a scheduling rule in the aircraft engine maintenance. Int. J. Prod. Econ. (2012). doi:10.1016/j.ijpe.2012.10.022 (Accepted for publication)
4. Reményi, C., Staudacher, S., Becker, H., Dinc, S., Manejev, R., Fichtelmann R.: Decentralized job-shop control for the maintenance of aircraft engines. In: Spath, D. (ed.) Proceedings of the 21st ICPR International Conference on Production Research, Stuttgart July 31–Aug 4, (2011)
5. Guide, V.D.R. Jr, Spencer, M.S.: Rough-cut capacity planning for remanufacturing firms. Prod Plan and Control **8**(3), 237–244 (1997)
6. Reményi, C., Staudacher, S., Holzheimer, N., Schulz, S.: Simulation of the maintenance process in an Aircraft Engine Maintenance Company. In: Duffie, N.A., DeVries, M.F. (eds.) Proceedings of the 44th CIRP Conference on Manufacturing Systems, June 1–3, Madison, (2011)
7. Wagner, M.: Modellbasierte Arbeitskräfteplanung für stochastische Instandhaltungsereignisse in der zivilen Luftfahrt. Dissertation, Technical University of Berlin (2009)
8. Eickemeyer, S.C., Borcherding, T., Schäfer, S., Nyhuis, P.: Validation of data fusion as a method for forecasting the regeneration workload for complex capital goods. Int. J. Prod. Eng. Res. Dev. (2013). doi:10.1007/s11740-013-0444-8
9. Eickemeyer, S.C., Borcherding, T., Nyhuis, P.: Information fusion as a means of forecasting expenditures for regenerating complex investment goods. Int. J. Mech. Ind. Eng. **6**, 179–182 (2012)
10. Hofstede, G.: The business of international business is culture. Int. Bus. Rev. **3**(1), 1–14 (1994)
11. House, R. J., Hanges, P. J., Javidan, M., Dorfman, P. W., Gupta, V.:.Culture, leadership, and organizations. Sage (2004)
12. Töpfer, A., Duchmann, C.: Der Einfluss der Landeskultur auf die Qualitätsorientierung der Mitarbeiter und Wettbewerbsvorteile in der internationalen Produktpolitik-Empirische Basis und kulturpsychologische Fundierung. In Herausforderungen der internationalen marktorientierten Unternehmensführung (pp. 43–75). Gabler (2011)
13. Kutschker, M., Schmid, S.: Internationales management. Oldenbourg Verlag (2008)

Machine Tool Thermal Errors Reduction for 5-axis Machining of Aircraft Parts

Jerzy Jedrzejewski and Wojciech Kwasny

Abstract Problems connected with the reduction of thermal errors in the controllable axes of 5-axis machining centres are discussed. A strategy for improving machine tool thermal properties is proposed. The conditions for the accurate holistic modelling, simulation and compensation of thermal errors in controllable rotary axes for aircraft parts machining purposes are described.

Keywords Machine tool · Thermal error · Reduction · Compensation

Introduction

The aerospace industry needs highly efficient 5-axis machine tools for precise machining. One of the major components of this efficiency is volumetric accuracy, which is mainly determined by the thermal error. The reduction and compensation of this error is intensively investigated by both research centres and the machine tool producers themselves [1–3].

The primary role in improving machine tool thermal properties and reducing errors is played by error identification through measurements [4, 5] and especially through the accurate modelling and numerical simulation of error behaviour in the normal machine tool operating conditions [6], taking into detailed account all the interactions.

J. Jedrzejewski (✉) · W. Kwasny
Institute of Production Engineering and Automation, Wroclaw University of Technology, Wybrzeze Wyspianskiego Str. 27, 50370 Wroclaw, Poland
e-mail: jerzy.jedrzejewski@pwr.wroc.pl

W. Kwasny
e-mail: wojciech.kwasny@pwr.wroc.pl

B. Denkena (ed.), *New Production Technologies in Aerospace Industry*,
Lecture Notes in Production Engineering, DOI: 10.1007/978-3-319-01964-2_25,

Therefore the authors have focused their research on acquiring knowledge to enable the holistic (integrated) modelling of the heat sources in the machine tool structure to most faithfully render the actual power loss generation and heat transfer processes taking place in the real machine tool and in its assemblies. Only then the temperatures, deformations and thermal errors predicted by the holistic model can be consistent with the ones occurring during machine tool operation, especially in nonstationary states, and can be used to effectively improve the thermal properties of machine tools, including 5-axis machining centres.

In order to achieve the highest possible effectiveness of improvement in thermal properties one should use a strategy and the latest experimental and theoretical knowledge, which is the focus of this paper. The presented models relating to controllable axes are the result of research conducted by the authors and their team for many years.

Strategy for Improving Thermal Properties of Machine Tools

The consistent improvement of machine tools is vital for the effective reduction of their thermal errors. The strategy for this improvement requires the latest theoretical and experimental knowledge. The latter should be based on possibly most precise measurements of temperature, power losses, deformations and thermal displacements. Also the effects of the machining environment and process, taking into account their complexity, should be studied.

This is needed in order to create accurate calculation and error models and to predict errors and precisely compensate them.

Some machine tool manufacturers (e.g. OKUMA) develop their own machine tool thermal improvement strategies and achieve good results. The more complex the machine tool structure, the more vital its consistent and comprehensive improvement based on a proper strategy.

Modelling of Thermal Errors

In the holistic thermal model of the 5-axis HSC centre shown in Fig. 1, it is particularly important to accurately model the bearing assemblies of the spindle and of the whole electrospindle since it is in these assemblies that the largest and highly variable heat sources are located. The contribution of this heat to the total thermal error occurring between the table and the spindle tip may exceed 75 %.

For this purpose a mathematical model of power loss generation in the bearings, taking into account all the interdependences between changes in temperature, thermal changes in the dimensions of the spindle/electrospindle assembly

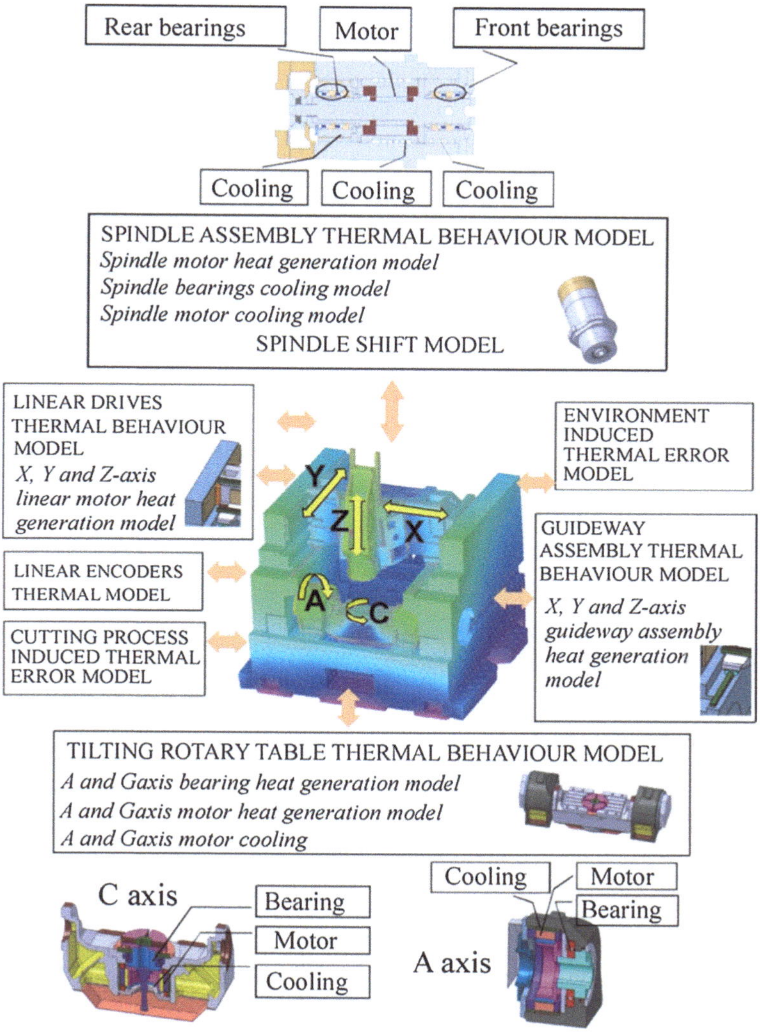

Fig. 1 Major heat sources in 5-axis vertical machining centre

components and the variable natural and forced heat transfer conditions, was developed and validated [7, 8]. Figure 1 shows, besides the thermal errors generated in the drives of rotary controllable axes Z, C and A, also the errors in linear axes X, Y and Z and the errors generated by the cutting process (the latter errors are not discussed in the present paper). In the case of the 5-axis machining of slender parts, e.g. aircraft engines, also the errors introduced by the production jig in the process of positioning, supporting and fixing the parts should be included in the proposed holistic error model. These are both thermal errors and errors generated by the forces exerted by the actuators.

A set of relations used in the modelling of power losses in controllable axes was presented in [6]. For the 5-axis HSC centre this model has been significantly refined in order to take into account the power loss generation and the heat transfer specific for the rotational spindle speeds as high as 50,000 rpm. For this purpose it was necessary to develop a shift model [9] for doing calculations on the basis of the predicted (for a given spindle speed) high-speed bearing contact angles which determine the forces acting on the rolling elements of the bearings. System SATO (Fig. 2), previously used to model mainly machine tool assemblies and simple structures of whole machine tools [10], has been developed by the authors for the purposes of the efficient holistic modelling of machine tool spindle assemblies and headstocks. The system is used to implement the holistic model shown in Fig. 1, in the part relating to the generation of power loss variation, taking into account the interdependences within the headstock structure and other rotational axes.

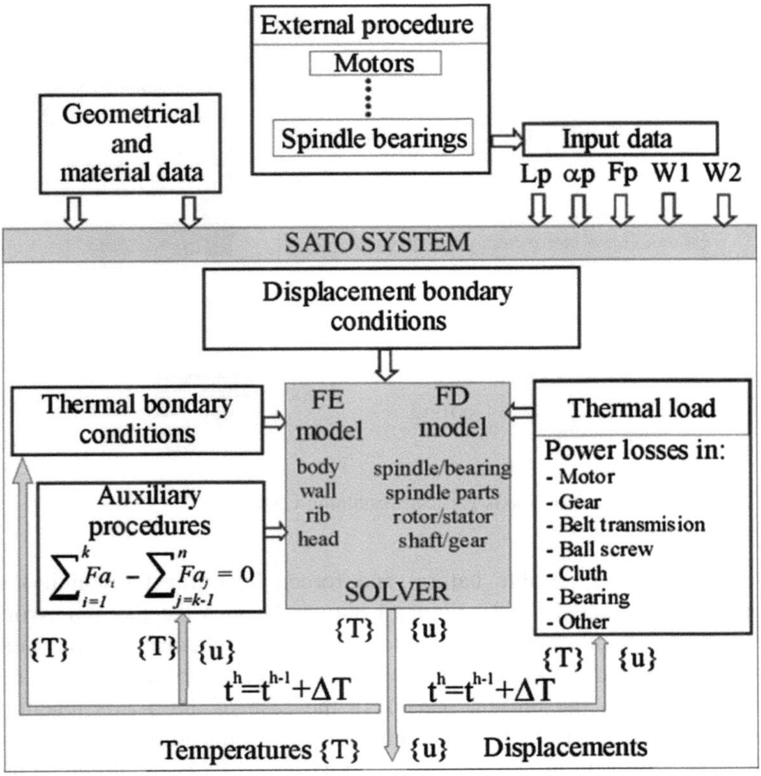

Fig. 2 General structure of SATO system

SATO includes detailed algorithms for determining power losses in spindle bearings, rolling, toothed and belt transmissions, motors and guides. It also comprises accurate heat transmission models and models of spindle and its bearing set behaviour under great centrifugal forces which occur at rotational speeds as high as 50,000 rpm.

The SATO structure incorporates the hybrid FEM and FDM modelling of spindle assemblies, mathematical models of the natural generation of power losses in heat sources of different kind, and detailed procedures specifying heat transfer depending on the geometric structures. As a result, the system enables the holistic modelling of the particular machine tool modules.

Also the tilting table motion drives are considerable sources of thermal errors in the 5-axis machining centre.

A general thermal model and an algorithm for computing temperatures and displacements in rotation axis A and C of the centre's tilting table are shown in Fig. 3 [11]. The algorithm is based on the catalogue specifications relating to

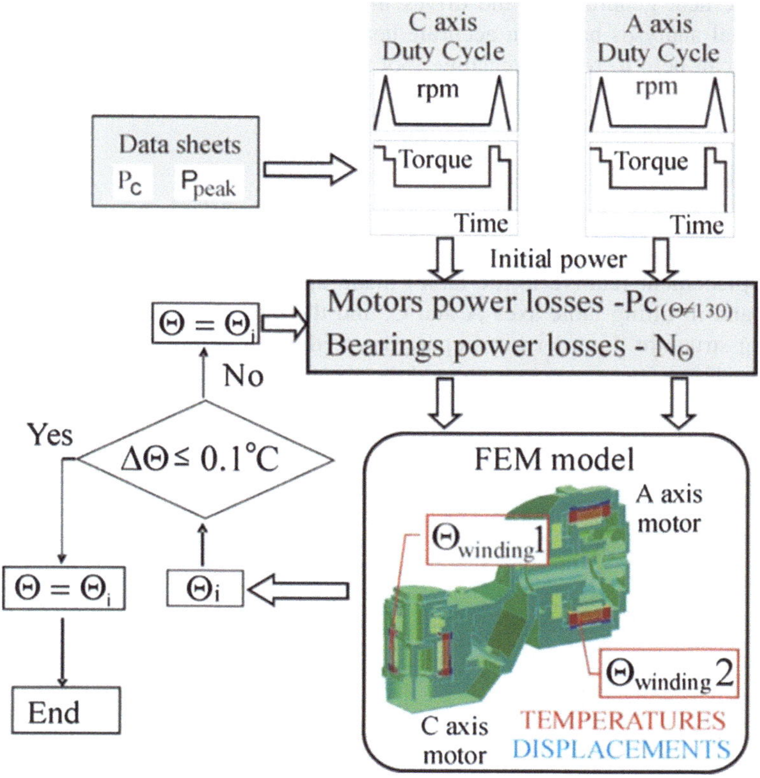

Fig. 3 Algorithm computing power losses and temperature in torque motors for temperature resolution and displacement estimation

direct high-torque motors, winding resistance–temperature dependencies, motor coolant heat transfer coefficients and the heat capacity and surface area of the table components transferring heat to the environment. Thermal errors are reduced mainly by reducing power losses and through the forced cooling of the closed spaces (within the range of heat source action) in housings and through the thermosymmetric machine tool design.

Power losses in high-speed (rolling bearing) friction nodes are reduced by maintaining a minimum lubricant film thickness enabling elastohydrodynamic lubrication using air-oil lubrication systems and the latest lubrication systems: MQGS (Minimum Quantity of Grease Supply) and MQCJ (Minimum Quantity and Cooling Jet) [12, 13]. The latter system combines lean lubrication of the bearing with its intensive cooling whereby very high high-speed coefficients, as high as DmN $= 5 \times 10^6$, can be used, which is unachievable in the case of other bearing lubrication methods.

A significant reduction in thermal errors can also be achieved by reducing to minimum the running time of the drives in accordance with the JIT rule quoted above. It is also essential to insulate the machine tool's load-bearing components from the heat generated in the drives in the course of machining. Undoubtedly numerical analyses based on accurate heat generation, distribution and exchange models can help to reduce machine tool thermal errors.

Scientific Result

The work on the holistic model led to the forecasting of the thermal error (spindle tip displacement relative to the table) made up of the errors arising in the headstock and in tilting table axes A and C, and the thermal elongations of the load-bearing structure. In order to compare the predicted thermal error values with the measured ones it was necessary to add the abrupt change in spindle position caused

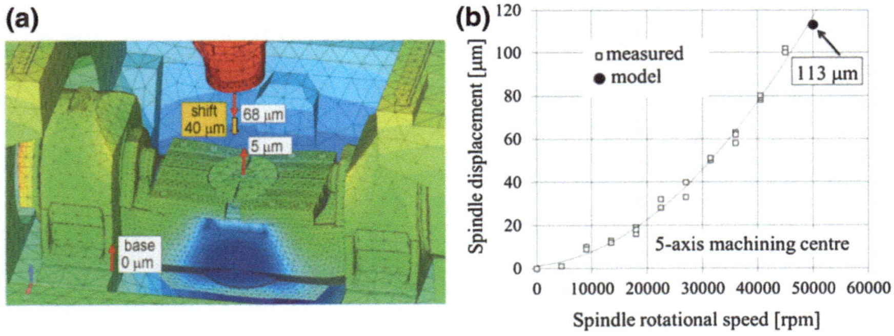

Fig. 4 Results of thermal deformation 5 axis machining centre and spindle shift calculations for 50,000 rpm (**a**), experimental verification of total spindle displacement relative to table in thermally steady states against displacements for other spindle rotational speeds (**b**)

by shift, predicted by the model described in [9], to the thermal error. For the example shown in Fig. 4 the error in direction Z, determined from the measurement curve amounts to about 116 μm after 3 h of work at 50,000 rpm while the one predicted using the holistic model amounts to 113 μm.

This means that the model satisfies the accuracy requirements for the simulation, analysis and improvement of the thermal properties of the precision 5-axis machining centres used in the aerospace industry.

Compensation of Thermal Errors

The compensation of thermal errors is undoubtedly the cheapest way of reducing their effect on machining precision. Hence it is the subject of intensive research focusing on the identification of the errors through measurements, and their modelling. In the case of spindle assemblies, axial thermal errors and shift add up. At a rotational speed of 50,000 rpm shift may exceed 35 % of the total spindle displacement. It is always the sum of the errors, one being the function of the machine tool thermal state (temperature) and the other being the function of the spindle rotational speed, which is measured. In order to develop an accurate compensation function the measured quantities need to be separated and the

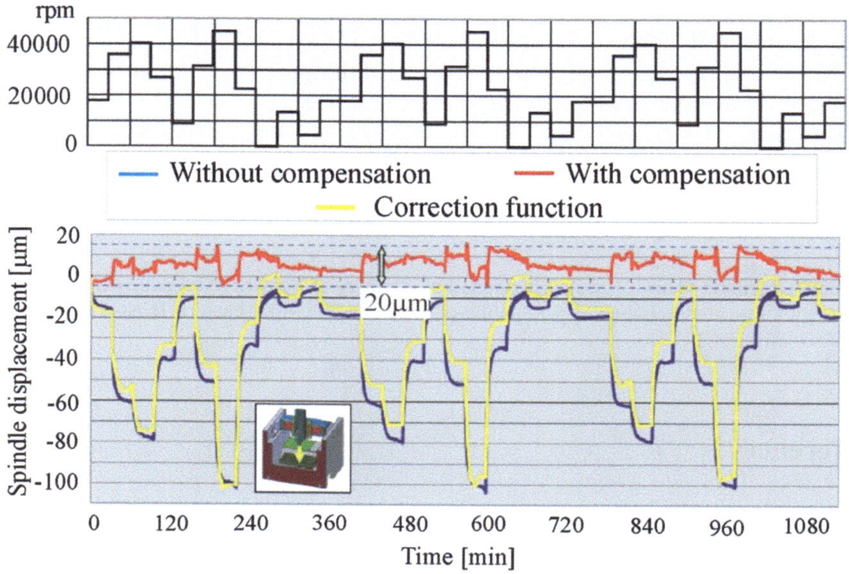

Fig. 5 Experimentally identified and predicted high-speed machining centre spindle displacements and theoretical compensation results

correction should be conditioned on two parameters: the thermal state and rotational speed of the spindle. An example of such compensation is shown in Fig. 5.

Figure 5 shows predictions of compensation in axis Z for a 5-axis centre with the maximum rotational spindle speed of 45,000 rpm. Each of the curves represents the sum of thermal displacement and shift, but because of the high rotational spindle speeds, the character of the curves is not exclusively thermal. The error after compensation does not exceed 20 % of the value measured before compensation (about 20 μm). At low rotational speeds (<20,000 rpm) the percentage of shift is small whereby the displacement curves have a character of typical thermal curves.

In order to obtain very small spindle displacement errors (<10 %) without compensation the machine tool assemblies need to be intensively cooled. It is also possible to take into account the thermal error component resulting from changes in ambient temperature, as an additional component of the compensation functions discussed above or through a separate compensation function (the leading machine tool manufacturers have already adopted the latter solution).

Conclusions

The modelling of thermal errors significantly contributes to the knowledge of their behaviour in the machine tool operating conditions. Errors in spindle rotation axis Z and tilting table axis C can be effectively modelled. Modelling errors in the whole machine tool workspace and the accurate experimental verification of such models are still difficult.

In the case of highly efficient high-speed milling centres, the accurate compensation of thermal errors still needs to be improved.

This paper has shown that it makes sense to holistically model the thermal errors of 5-axis HSC centres which in order to machine aircraft parts must meet rigorous precision requirements. The holistic model needs to be fine-tuned and extended to cover all the components of the HSC centre thermal error.

Acknowledgments This research is supported by Ministry of Science and Education of Poland, Ministry of Knowledge Economy—Republic of Korea and NCBiR (Project: INNOTECH—K1/IN1/75/155671/NCBR/13).

References

1. Abele, E., Altintas, Y., Brecher, C.: Machine tool spindle units. CIRP Ann. Manufact. Technol. **59**(2), 781–802 (2010)
2. Mayr, J., Jedrzejewski, J., Uhlmann, E., et al.: Thermal issues in machine tools. CIRP Ann. Manufact. Technol. **61**(2), 771–791 (2012)

3. Turek, P., Jedrzejewski, J., Modrzycki, W.: Methods of machine tool error compensation. J. Mach. Eng. **10**(4), 5–26 (2010)
4. Kwasny, W., Turek, P., Jedrzejewski, J.: Survey of machine tool error measuring methods. J. Mach. Eng. **11**(4), 7–38 (2011)
5. Schwenke, H., Knapp, W., Haitjema, H., et al.: Geometric error measurement and compensation of machines—an update. Ann. CIRP Ann. Manufact. Technol. **57**(2), 660–675 (2008)
6. Jedrzejewski, J., Kwasny, W.: Knowledge base and assumptions for holistic modelling aimed at reducing axial errors of complex machine tools. J. Mach. Eng. **13**(2), 7–25 (2013)
7. Jedrzejewski, J.: Thermal problems in machine tools design and operation. In: Mekid, S. (ed.) Introduction to precision machine design and error assessment, pp. 75–127. CRC Press, (2009)
8. Jedrzejewski, J., Kowal, Z., Kwasny, W., Modrzycki, W.: Hybrid model of high speed machining centre headstock. CIRP Ann. Manufact. Technol. **53**(1), 285–288 (2004)
9. Jedrzejewski, J., Kwasny, W.: Modelling of angular contact ball bearings and axial displacements for high-speed spindles. CIRP Ann. Manufact. Technol. **59**(1), 377–382 (2010)
10. Winiarski, Z., Kowal, Z., Kwasny, W., Ha, J.-Y.: Thermal model of the spindle drive structure. J. Mach. Eng. **10**(4), 41–52 (2010)
11. Blazejewski, A., Kwasny, W., Jedrzejewski, J., Gim, T.-W.: Modelling thermal deformation of tilting rotary table with direct drive system. J. Mach. Eng. **10**(4), 26–40 (2010)
12. Koyama, M.: Minimum quantity and cooling jet lubricated angular contact ball bearings for machine tool. NTN Tech. Rev. **74**, 24–27 (2006)
13. Lee, S.-W., Maeda, T.: Angular contact ball bearings that require a minimum quantity of grease supply for lubrication. NTN Tech. Rev. **75**, 90–95 (2008)

Recycling of Aluminum Chips by Hot Extrusion

Matthias Haase, Andreas Jäger and A. Erman Tekkaya

Introduction

The production of primary aluminum by processing bauxite is one of the most energy intense processes of metal production. The energy required for the production of one ton of aluminum (168 GJ/t) requires almost ten times the energy required for the production of one ton of steel (20 GJ/t) [5, 6]. By using secondary aluminum, produced by recycling aluminum scrap, the energy requirement can be reduced by up to 94 % [2]. Considering the development of the global aluminum production (see Table 1), the use of secondary aluminum has a high potential for saving energy and therefore energy costs.

One approach of further reducing the energy demand for the production of aluminum products is the direct recycling of aluminum scrap by forming processes. Here, the energy intense melting process for the production of secondary aluminum can be avoided. One example of this approach is the direct hot extrusion of aluminum machining chips.

Hot Extrusion of Aluminum Machining Chips

The conventional process of hot extrusion is used for the production of semi-finished aluminum profiles (extrudates). The shape of the extrudates can reach from simple bars to complex multi-hollow geometries [1]. The basic process principle of hot extrusion is shown in Fig. 1.

In conventional hot extrusion, a preheated billet (350–550 °C) is inserted into the container of the extrusion press and pushed by a moving ram through the die

M. Haase (✉) · A. Jäger · A. E. Tekkaya
Technische Universität Dortmund, Institut für Umformtechnik und Leichtbau,
Baroper Straße 301, 44227 Dortmund, Germany
e-mail: Matthias.Haase@iul.tu-dortmund.de

B. Denkena (ed.), *New Production Technologies in Aerospace Industry*,
Lecture Notes in Production Engineering, DOI: 10.1007/978-3-319-01964-2_26,

Table 1 Development of the global aluminum production [3, 7]

Aluminum production	1970	1980	1990	2000	2010	Est. 2030
Primary (in Mio t)	10.3	16.1	19.4	24.5	41.2	26–28
Secondary (in Mio t)	2.3	3.7	6.1	8.4	8.4	22–24

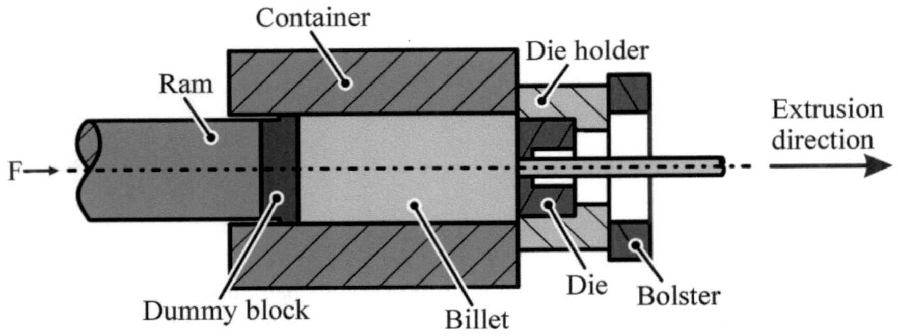

Fig. 1 Process principle of hot extrusion

Fig. 2 Process principle of hot extrudates [4]

orifice, which determines the geometry of the extrudate. For the extrusion of aluminum machining chips, the basic process principle is extended by the processing of the chips, as shown in Fig. 2.

In the first step, the machining chips are cleaned from remaining lubricants by thermal or chemical treatment. In the second step, the chips are compacted on a hydraulic press to chip-based billets usable for hot extrusion. To homogenize the material, the chip-based billets are preheated in a third step prior to extrusion. Finally, the chip-based billets are hot extruded to produce chip-based extrudates.

Results

To evaluate the mechanical properties of chip-based extrudates, lubricant-free AA6060 aluminum alloy machining chips were compacted and hot extruded. As a reference, conventional cast billets of the same alloy were also extruded. The chips

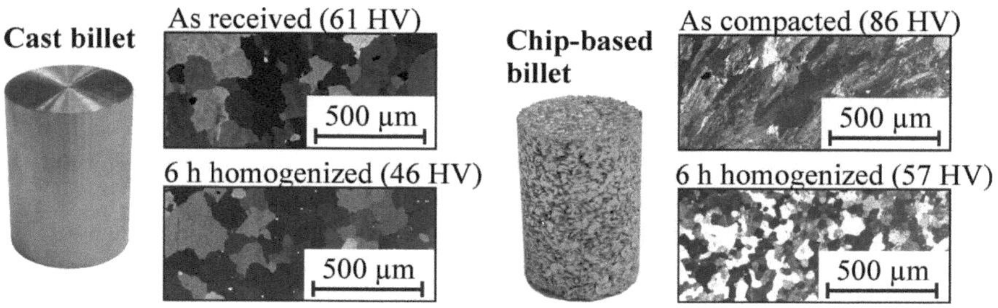

Fig. 3 Microstructure and hardness of chip-based and cast billets before and after homogenization

were compacted with a force of 500 kN in a steel tube with an inner diameter of 60 mm to chip-based billets. The density of the chip-based billets was 80 % compared to the density of cast aluminum. The chip-based and the cast billets were homogenized for 6 h at 550 °C in an electrical furnace prior to the extrusion. Figure 3 shows the resulting microstructure and the measured hardness values of the billets before and after homogenization.

The microstructure of the cast billets did not significantly change due to the homogenization. However, the microstructure of the chip-based billet changes from severely deformed grains to homogeneous fine grains after homogenization. This can be related to recrystallization processes initiated by the high amount of strain imposed on the material during machining. As the microstructure strongly influences the mechanical properties of the material, the *Vickers* hardness was measured before and after homogenization of the billets. Before homogenization, the hardness of chip-based extrudates is higher compared to the hardness of the cast billets, which can be related to the work hardening of the chips during machining. After homogenization, the hardness of the chip-based billets remains higher compared to the hardness of the cast billets, which is due to the smaller average grain size of the chip-based billet.

After homogenization, the chip-based and cast billets were extruded on a 2.5 MN direct extrusion press (Collin LPA250t) at 550 °C to solid rectangular profiles with a 20 mm × 5 mm geometry. The extrusion ratio, the ratio between the cross section of the extrudate and the cross section of the container of the extrusion press, was $R = 34$. The ram speed was kept constant at 1 mm/s. Three different dies were used for the extrusion trials. The first die is a flat-face die conventionally used for the extrusion of solid section profiles. The second die is a porthole die, which is conventionally used to produce hollow profiles. By removing the mandrel, the die can be used for the production of solid sections. The third die is an experimental extrusion die based on a concept of Estrin et al. [2], which integrates the process of ECAP into the process of hot extrusion. In this integrated extrusion and ECAP die (iECAP die), the material is pressed around defined angles inside the die. The last two die concepts aim on increasing the shear and pressure affecting the chip-based billets and therefore to increase the welding quality between the individual chips. The three tools are shown in Fig. 4.

Independent of the die concept, the extrusion of chip-based billets led in all cases to profiles without visible surface defects. To investigate the mechanical properties of chip-based extrudates and extruded cast billets depending on the extrusion die, tensile tests were conducted (see Fig. 5).

Four tensile tests for each alignment were conducted in accordance with EN ISO 6892-1:2009 at room temperature with a constant strain rate of $2.5 \times 10^{-3}\ s^{-1}$. The standard deviation for chip-based and cast billets was in all cases below 12 and 7 MPa for the ultimate tensile strength and below 4 and 2 % for the uniform elongation, respectively. The results show that the tensile strength of extruded cast billets is, depending on the extrusion die, up to 12 % higher compared to the tensile strength of chip-based extrudates. For the chip-based extrudates, the highest tensile strength is achieved by using the iECAP die. It is evident that the ductility of the extrudates depends more significantly on the die used for the extrusion compared to the tensile strength. For the extruded cast billets, an increase in uniform elongation of 75 and 99 % was achieved by changing the extrusion die from the flat-face die to the porthole die and to the iECAP die, respectively. For the chip-based extrudates, the increase in uniform elongation was 97 and 131 %, respectively.

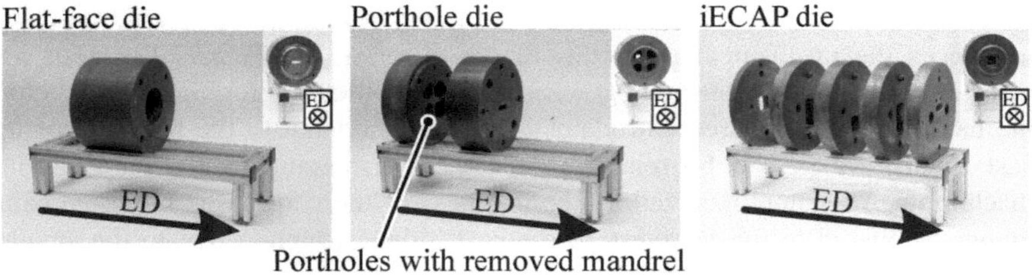

Fig. 4 Three different die concepts for chip extrusion

Fig. 5 Tensile test results for different extrusion dies

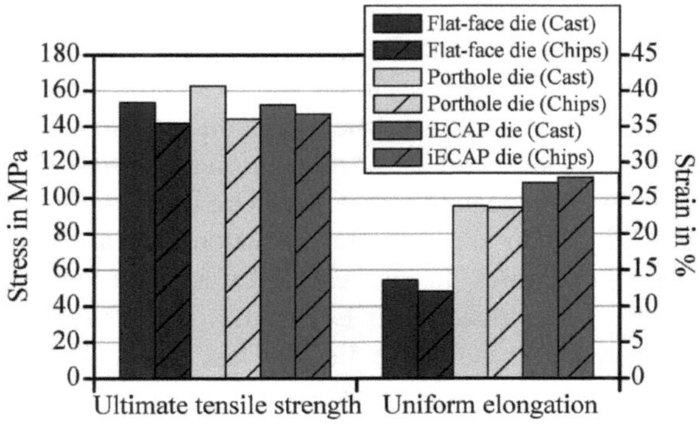

Conclusion and Outlook

The direct recycling of aluminum machining chips by hot extrusion is a promising approach to increase the energy efficiency of aluminum recycling. With an appropriate die design, it is possible to produce chip-based extrudates with mechanical properties comparable to extruded cast billets. The ductility of the extrudates depends on the extrusion die design, while their strength was not significantly affected. In future works, the production of finished products by cold extrusion of the chip-based extrudates is investigated.

Acknowledgments The authors thank the Graduate School of Energy Efficient Production and Logistics at the Technische Universität Dortmund for the support of this research.

References

1. Bauser, M.: Metallkundliche Grundlagen. In: Strangpressen. 2. Auflage, Aluminium-Verlag, Düsseldorf (2001)
2. Estrin, Y., Ferkel, H., Hellmig, R.J., Lamark, T., Popov, M.V.: German Patent DE102005049369 (2008)
3. GDA.: Gesamtverband der Aluminiumindustrie e.V. http://www.aluinfo.de (2013). Stand: 4 Jun 2013
4. Haase, M., Ben Khalifa, N., Tekkaya, A.E., Misiolek, W.Z.: Improving mechanical properties of chip-based aluminium extrutes by integrated extrusion and equal channel angular pressing (iECAO). Mater. Sci. Eng., A. **539**, 194–204 (2012)
5. Price, L., Sinton, J., Worrell, E., Phylipsen, D., Hu, X., Li, J.: Energy use and carbon dioxide emissions from steel production in China. Energy **27**, 429–446 (2002)
6. Schmitz, C.: Handbook of Aluminum Recycling, pp. 27–30. Vulkan Verlag, Essen (2006)
7. Schwarz, H.G.: Aluminium production and energy. Encyclopedia of Energy **1**, 81–95

